Communications
in Computer and Information Science 156

David Obdržálek   Achim Gottscheber (Eds.)

Research and Education in Robotics - EUROBOT 2010

International Conference
Rapperswil-Jona, Switzerland, May 27-30, 2010
Revised Selected Papers

 Springer

Volume Editors

David Obdržálek
Charles University in Prague
Faculty of Mathematics and Physics
Malostranské náměstí 25
118 00 Praha 1, Czech Republic
E-mail: david.obdrzalek@mff.cuni.cz

Achim Gottscheber
SRH University Heidelberg
69121 Heidelberg, Germany
E-mail: achim.gottscheber@fh-heidelberg.de

ISSN 1865-0929   e-ISSN 1865-0937
ISBN 978-3-642-27271-4   ISBN 978-3-642-27272-1 (eBook)
DOI 10.1007/978-3-642-27272-1
Springer Heidelberg Dordrecht London New York

Library of Congress Control Number: 2011943226

CR Subject Classification (1998): I.2.9, I.2, I.4, I.5, H.4, K.3, I.6

Typesetting: Camera-ready by author, data conversion by Scientific Publishing Services, Chennai, India

Printed on acid-free paper

Springer is part of Springer Science+Business Media (www.springer.com)

Preface

This volume contains the selected papers presented at the International Conference on Research and Education in Robotics—EUROBOT 2010—held in Rapperswil-Jona, Switzerland, May 27–30, 2010. The Conference was organized as part of the annual EUROBOT event.

A fundamental aspect of EUROBOT is promotion of science and technology among young students and researchers. Therefore, the conference was accompanied by the EUROBOT Autonomous Robot Contest (specifically its international finales): three best teams from each participating country brought their robots to the city of Rapperswil-Jona in Switzerland to meet their colleagues from all over the world—teams from 24 countries took part in 2010.

Besides the conference presentations, two invited talks were given by:

- Bruno Siciliano, Director of Prisma Lab, University of Naples, Italy
- Veronique Raoul, President of EUROBOT Association

The conference was supported by the EUROBOT Association, the town of Rapperswil-Jona, Switzerland, the HSR University of Applied Sciencies Rapperswil, Switzerland, and the Charles University in Prague, Czech Republic.

We would like to thank everyone involved in the conference organization and in the whole EUROBOT event in Rapperswil-Jona. We also thank all the authors who submitted their work and the reviewers who helped select the best papers to be presented and be included in these proceedings. Last, but not least, we want to thank the developers of the EasyChair system, which was used as the main management tool for the conference papers.

May 2010

David Obdržálek
Achim Gottscheber

Organization

EUROBOT 2010 was organized by:

- EUROBOT Association
- SRH University Heidelberg, Germany
- Charles University in Prague, Czech Republic
- Hochschule für Technik Rapperswil, HSR/IMA, Rapperswil, Switzerland in cooperation with Planète Sciences Association, France.

Executive Committee

Conference Chair

Achim Gottscheber   SRH University Heidelberg, Germany

Program Chair

David Obdržálek   Charles University in Prague, Czech Republic

Local Organization

Agathe Koller   Hochschule für Technik Rapperswil, HSR/IMA,
                           Rapperswil, Switzerland

Program Committee

Jacques Bally   Y-Parc, Parc Scientifique et Technologique,
                           Yverdon-les-Bains, Switzerland
Kay Erik Böhnke   SRH University Heidelberg, Germany
Jean-Daniel Dessimoz   Western Switzerland University of Applied
                           Sciences, HESSO HEIG-VD,
                           Yverdon-les-Bains, Switzerland
Boualem Kazed   University of Blida, Algeria
Agathe Koller   Hochschule für Technik Rapperswil, HSR/IMA,
                           Rapperswil, Switzerland
Markus Kottmann   Hochschule für Technik Rapperswil, HSR/IMA,
                           Rapperswil, Switzerland
David Obdržálek   Charles University in Prague, Czech Republic
Julio Pastor Mendoza   Universidad de Alcalá, Madrid, Spain

Table of Contents

Improved Image Dominant Plane Extraction for Robot Navigation 1
Andrés Colín-Espinoza, Héctor Alejandro Montes-Venegas, and
María Enriqueta Barilla-Pérez

Do Cognition and Robotics Share Common Ground? Some Answers
from the MCS Perspective and an RH-Y Case Study 14
Jean-Daniel Dessimoz

Elements of Hybrid Control in Autonomous Systems and Cognitics 30
Jean-Daniel Dessimoz

Mechanical Design and System Architecture of a Tracked Vehicle
Robot for Urban Search and Rescue Operations 46
Raimund Edlinger, Andreas Pölzleithner, and Michael Zauner

Flexible Robot Strategy Design Using Belief-Desire-Intention Model 57
Loris Fichera, Daniele Marletta, Vincenzo Nicosia, and
Corrado Santoro

Using Quadtrees for Realtime Pathfinding in Indoor Environments 72
Julian Hirt, Dominik Gauggel, Jens Hensler, Michael Blaich, and
Oliver Bittel

Robot Workshop and Contest for High-School Students Organized by
"Politehnica" University of Bucharest 79
Sanda Paturca, Catalina Enescu, Constantin Ilas, and
Alexandru Morega

Requirements and Solutions in Applied Robotics 87
Rahman Jamal

Estimation of Mobile Robot Pose from Optical Mouses 93
Lenka Mudrová, Jan Faigl, Jaroslav Halgašík, and Tomáš Krajník

Performance Comparison of Vision Sensors and Processing Power of
Two Robotic Platforms for Obstacle Avoidance 108
Sanda Paturca, Dan Novischi, and Constantin Ilas

Combining Gaussian Processes and Conventional Path Planning in a
Learning from Demonstration Framework 118
Markus Schneider, Richard Cubek, Tobias Fromm, and
Wolfgang Ertel

Development, Realization and Control of a Mobile Robot 130
    Ralf Stetter, Paweł Ziemniak, and Andreas Paczynski

Startup Robotics Course for Elementary School 141
    Dmitry Sukhotskiy and Anton Yudin

Micromouse – Electronics on Wheels 149
    Tony Wilcox

Autonomous Mobile Robot Development in a Team, Summarizing Our
Approaches 168
    Andrey Demidov, Andrey Kuturov, Anton Yudin,
    Boris Krasnobryzhiy, Mikhail Chistyakov, and
    Rustam Borovik

Distributed Control System in Mobile Robot Application:
General Approach, Realization and Usage 180
    Andrey Vlasov and Anton Yudin

A Modular and Scalable Electronic System for Mobile and Autonomous
Robots 193
    Raimund Edlinger and Michael Zauner

Author Index 201

Improved Image Dominant Plane Extraction for Robot Navigation

Andrés Colín-Espinoza*, Héctor Alejandro Montes-Venegas, and María Enriqueta Barilla-Pérez

Universidad Autónoma del Estado de México
Toluca, Estado de México, México
nothingsburning@gmail.com,
{h.a.montes,barillam}@fi.uaemex.mx

Abstract. The dominant plane is a planar area that occupies the largest portion of an image captured by a camera. In this paper, we present an improved method for extracting the dominant plane from optical flow and use the resultant image to steer a mobile robot along a corridor route. In this route the dominant plane is the free space where the robot can navigate without colliding with the obstacles cluttering the environment. Our experimental results illustrate significant performance improvements over those published in previous works.

1 Introduction

Visual navigation is one of the most challenging problems in robotics. Vision is perhaps the most powerful sense but also the most complicated. The main purpose of a robotic vision system is to determine appropriate control movements using the images captured by a camera mounted on the robot.

There exist a wealth of methods and techniques to extract the information needed by a robot that has a vision system [4]. Some robots use previously constructed maps to carry out their navigation [5][13]. Others construct their own map from the environment [9][12] and some robots even navigate without map [11]. In this paper we will focus on mapless navigation using optical flow. Optical flow is an estimation of image motion defined as the projection of velocities of $3D$ surface points onto the imaging plane of a visual sensor [2].

In a navigation system that uses optical flow, the system is able to detect objects with or without movement as the robot moves [7], track objects for target pursuit [8] and perform motion segmentation of objects [14], among many other tasks. By using optical flow we can separate objects with different depths from the camera that are finite or infinite with respect to the observer. In this paper we present an improved algorithm for the extraction of the dominant plane from the optical flow in two successive images. The dominant plane is defined as the largest region on the image [9][10]. This region is the free space where a mobile robot can navigate without colliding with the obstacles present in its workspace.

---

* This work was supported by the UAEMex under project grant number 2570/2007U.

D. Obdržálek and A. Gottscheber (Eds.): EUROBOT 2010, CCIS 156, pp. 1–13, 2011.

We begin in Section 2 by covering the basics of the optical flow estimation. This is followed in Section 3 by a detailed description of our method for the extraction of the dominant plane. In Section 4 we discuss the results of our experiments. Section 5 concludes the paper.

2   Visual Navigation

The procedure to determine the free space where a robot can move begins with a sequence of ordered images captured by a camera mounted on a mobile robot moving along a corridor path. These sequences of images allow the estimation of motion as discrete image displacements. We first estimate the optical flow by computing a vector flow field of the scene. We then proceed to compute the dominant plane by selecting three random points from the vector flow field, with these points we compute the affine coefficients needed to obtain the planar flow. To extract the dominant plane a discrimination criterion is computed from the arithmetic mean of the lengths of the optical flow vectors called *tolerance range*. This range is compared with the difference between the optical flow vectors and the planar flow vectors. For each vector, if the difference is less than the tolerance range, then the vector belongs to the dominant plane.

2.1   Optical Flow

Optical flow is the estimation of apparent movement of a 3D projection onto the image plane caused by the relative motion between the camera and the scene, describing the direction and velocity of the characteristics inside the image [2]. This movement is represented with a vector field, where each vector is the

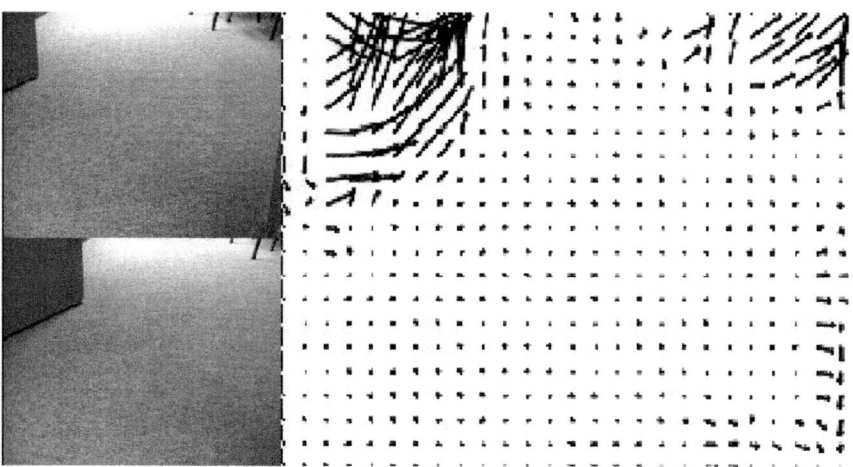

Fig. 1. Optical flow of two successive images represented by a vector field shown right

displacement of a point in a sequence of images obtained by a moving camera. Figure 1 shows the vector field of an obstacle in different locations of the scene.

In order to calculate the optical flow of successive image frames, we need to find the brightness value of a point from the first image and again in the next image. The distance from the point in the first image to the point in the second image is the optical flow vector of that point, and the collection of all displacement vectors of brightness values between the two images is the optical flow.

Formally, the optical flow vector is defined as the displacement $\mathbf{d} = (d_x, d_y)$ that occurs in a point $\mathbf{u} = (u_x, u_y)$ from an image $\mathbf{I}$ to an image $\mathbf{J}$, and its new position $\mathbf{J}$ is defined as $\mathbf{v} = (u_x + d_x, u_y + d_y)$.

There are many methods to estimate the optical flow [1][6], and each method depends on a number of assumptions, restrictions and constraints. A performance study of many of these methods can be found in [1]. In this paper we use the pyramidal Lucas-Kanade method [3] to compute the optical flow from an image J. To find the optical flow, the flow vectors need to minimize the equation 1:

$$J_{LK}(w) = \int \int_{\Omega(x)} w^2(x) |\mathbf{u}^T \nabla \, f(x,y,t) + f_t|^2 dx \tag{1}$$

Where $\Omega(x)$ is the integration window centered at x, and w is the weight function (pixel values). $\mathbf{u} = (\frac{dx}{dt}, \frac{dy}{dt})^T$ and $f(x,y,t)$ is the image at time t.

The main idea behind this method is to find the optical flow in recursive form, where the pyramidal representation of images help to track large displacements of objects in the sequence. The method consists of computing the optical flow at the top level of the pyramid and propagating the resultant flow to the immediate inferior level of the pyramid until the original image is reached. Figure 2, shows the pyramidal representation of the Lucas-Kanade method. The method allows the detection of large movements of pixels in the sequence of images [3].

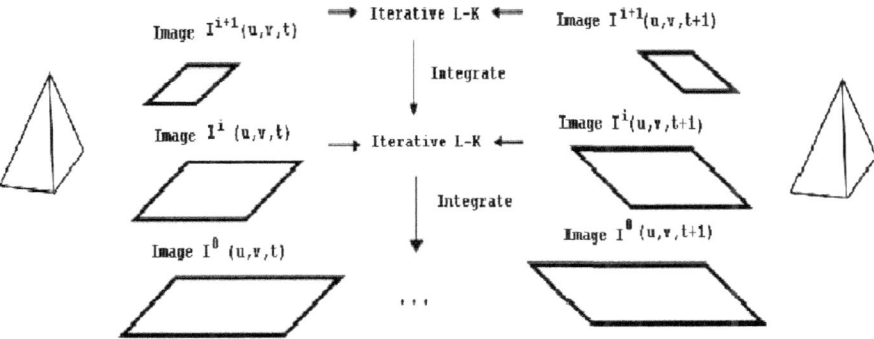

Fig. 2. The Lucas-Kanade method with pyramids, where the optical flow is calculated at the top of the image and integrates the resultant flow to the next inferior level of the pyramid until the original image size is reached

2.2 Dominant Plane Detection

Once the optical flow is found, we proceed to calculate the dominant plane, defined as the area that occupies the largest domain in the observed image. This area is the free space in the scene where a mobile robot can navigate. In order to obtain the detailed form of the objects in the space observed, the distance of the camera to the plane must be finite in a space [9][10].

Figure 3 shows an obstacle that occupies only a small part in the image, while the floor occupies almost all the image and is considered the dominant plane.

For computing the dominant plane, we need to make a discrimination between the movement of the objects determined by the optical flow and the movement of the scene determined by the planar flow. The planar flow is the vector field that represents the motion flow of the largest area in the image sequence [10]. As in optical flow, to estimate de planar flow, we need to find the same brightness value from a first image and again in the next image. In this case the use of affine transformation is needed to obtain the displacement vectors.

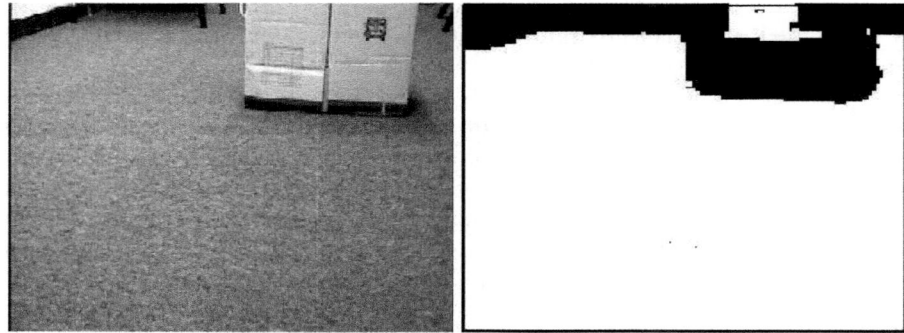

Fig. 3. In the left image the floor is the largest area. In the right image the white area is the dominant plane of the left image, and the black area is considered to be the obstacles in the scene.

In a pair of consecutive images, the projection of the dominant plane movement is estimated by the equation 2:

$$\begin{pmatrix} u' \\ v' \end{pmatrix} = \begin{pmatrix} a & b \\ d & e \end{pmatrix} \begin{pmatrix} u \\ v \end{pmatrix} + \begin{pmatrix} c \\ d \end{pmatrix} \qquad (2)$$

Small displacement by the camera implies that the homography between the two images of a flat surface is approximated by one affine transformation that connects the corresponding points $(u, v)^T$ and $(u', v')^T$ [9][10]. From 2 the variables a,b,c,d,e,f are known as the affine coefficients, and with them it is possible to find the point $p' = (u', v')^T$ in the second image from the point $p = (u, v)^T$ with the homography equation 3 between the two images of a flat surface [9][10].

$$p = Hp'$$ (3)

where $p = (u, v, 1)^T$ and $p' = (u', v', 1)^T$ are corresponding points in the two images. The matrix H is expressed as the equation 4:

$$H = K(R + tn^T)K^{-1})$$ (4)

The internal parameters of the camera are as follows. R is the rotation matrix, t is the translation vector, n is the cross flow plane to the flat surface, K and K^{-1} are the affine transformations.

The camera parameters R and tn^T are approximated by an affine transformation when the camera displacement is small [9][10]. Then the planar flow is estimated from the corresponding points $p' = (u', v')^T$ and $p = (u, v)^T$.

Once we have found the optical flow and the planar flow of two consecutive images, the estimation of the dominant plane is made by making a discrimination between the optical flow vectors and the planar flow vectors [10]. The discrimination of points that belong to the dominant plane area are those that do not change abruptly or have a longer extension from the rest of the vectors.

3 Improving the Dominant Plane Extraction

The method to compute the dominant plane as described in [9][10] uses several values defined experimentally. This paper proposes a dynamic determination of these values and a more adaptable algorithm for the resultant optical flow each time the dominant plane is computed.

The procedure we use for detecting the dominant plane is straightforward. First, we apply a bilateral filter to reduce noise and to preserve object edges on the sequence of the captured image. This filter operates in the image's range pixel values by replacing them with a weighted average of its neighbors in both space an intensity. Then the optical flow is calculated in two consecutive images as described above. Once we have the optical flow, the planar flow vectors are calculated and finally a discrimination of vectors must satisfy a predefined tolerance range to determine the dominant plane.

Because the points in the dominant plane are determined by the difference of the flow vectors between the optical flow and planar flow field, the tolerance factor of this difference is a chief component for deciding which points belong to the dominant plane and which do not. This difference is determined by a tolerance range that, most of the time, depends on the optical flow field. Given that the magnitude of the optical flow vectors is variant from one image to another, and this variability is related to the velocity of the robot, vibrations on the camera, noise in the images and changes in the illumination greatly affect the extraction of the dominant plane. Preserving sharp edges in the image sequence and dynamically computing the value of the adaptable tolerance range makes the entire procedure more robust and increases the accuracy of the extracted dominant plane.

After all this, we estimate the value of the tolerance range by an arithmetic mean of the optical flow vectors in two consecutive images. This provides adaptable values for the extraction of the dominant plane. By making this value dynamic, the algorithm for detecting the dominant plane adapts itself to these variable and often problematic situations. Finally, after finding the optical flow we proceed to compute the planar flow.

3.1   Planar Flow Estimation

For the estimation of the planar flow we select three non-collinear random points from the computed optical flow field. These points are needed to compute the affine coefficients. The points must belong to the free area because selecting at least one point that belongs to an obstacle would give an incorrect solution in the estimation of the affine coefficients and will result in a incorrect dominant plane [10].

Knowing that the optical flow is the dense correspondence of points between a pair of images, by applying the equation 2 to each point in the dominant plane, we can compute the planar flow using equation 5 [10]. If $(u, v)^T$ and $(u', v')^T$ are corresponding points in the two consecutive images, the point $(u', v')^T$ can be computed as follows:

$$\begin{pmatrix} u' \\ v' \end{pmatrix} = \begin{pmatrix} u \\ v \end{pmatrix} + \begin{pmatrix} \dot{u} \\ \dot{v} \end{pmatrix} \tag{5}$$

where $(\dot{u}, \dot{v})$ is the optical flow of the point (u, v).

With the three non-collinear points from the optical flow, equation 2 has a unique solution to compute the affine coefficients. Once these coefficients are calculated it is possible to detect the movement of the dominant plane in the sequence of images. This is the collection of points defined by equation 6 that represents the planar flow.

$$\begin{pmatrix} \hat{u} \\ \hat{v} \end{pmatrix} = \begin{pmatrix} u' \\ v' \end{pmatrix} - \begin{pmatrix} u \\ v \end{pmatrix} \tag{6}$$

If the dominant plane has already been calculated, then the affine coefficients for computing the planar flow in the next pair of images can be estimated with the least-squares method. Note that small displacements on the images produce insignificant changes on the dominant plane [9]. This means that the changes of the values of the affine coefficients are also negligible. To compute the new values of the affine coefficients, the points $(u_i, v_i)^T$ and $(u'_i, v'_i)^T$ where $(0 \leq i \leq n)$, are considered corresponding points in two consecutive images then the mean-squared error E_u and E_v defined by equation 2 are,

$$\begin{aligned} E_u &= \sum_{i=1}^{n} \{u'_i - (au_i + bv_i + c)\}^2, \\ E_v &= \sum_{i=1}^{n} \{v'_i - (du_i + ev_i + f)\}^2, \end{aligned} \tag{7}$$

In this case n is the number of points taken from the images to estimate the new values. We use the same random points used in the previous estimation of dominant plane for computing the affine coefficients that minimize E_u and E_v. Once the planar flow is calculated we determine the tolerance range and extract the dominant plane.

3.2   Tolerance Range ε

Because the estimation of the dominant plane greatly depends on the magnitude of the vectors of the optical flow, we dynamically determine the tolerance range ε to make the estimation of the dominant plane adaptable to the variations of the sequence of images. When small values of ε are experimentally set the iterative process in the estimation of the affine coefficients becomes a very time consuming process. Given that the difference of the values of optical flow and planar flow vectors are not constant in each pair of consecutive images, due to the different displacements that the robot can have when acquiring the images, we determine the value of ε by computing the arithmetic mean of the lengths of the vectors of the optical flow. This rather simple strategy makes this value variable and adaptable to abrupt changes in the images, thus increasing the accuracy and the safety of the navigation.

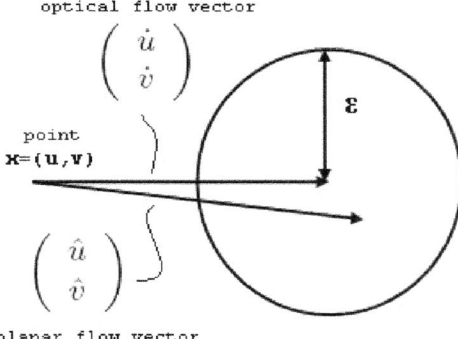

Fig. 4. The tolerance range is defined as the area where the optical flow vectors and the planar flow vectors must be inside the area in order to establish the point as a part of the dominant plane

The arithmetic mean of lengths from the optical flow vectors is simply defined as:

$$\varepsilon = \frac{1}{n} \sum_{1}^{n} \parallel (\dot{u}, \dot{v}) \parallel \tag{8}$$

Where n is the number of vectors obtained in the optical flow $(\dot{u}, \dot{v})$. The next step is to estimate the dominant plane with this value of ε.

3.3   Dominant Plane Detection

After computing the optical flow, planar flow and the tolerance range ε, then we proceed with the discrimination of the optical flow and planar flow vectors using equation 9:

$$\left| \begin{pmatrix} \dot{u} \\ \dot{v} \end{pmatrix} - \begin{pmatrix} \hat{u} \\ \hat{v} \end{pmatrix} \right| < \varepsilon \tag{9}$$

If the inequality is satisfied, then point $(u_t, v_t)^T$ is set as part of the dominant plane. Once the dominant plane is completed, if it covers at least 50% of the scene, it is said to be satisfactory [10]. Otherwise we need to recompute the affine coefficients to estimate the planar flow and also to recompute the dominant plane.

Once the dominant plane is extracted, we can give the appropriate commands to the robot's controller to perform a safe navigation.

4   Experimental Results

In this section we present a series of experimental results using different sets of image sequences. First, we determine a number of parameters necessary for computing the optical flow using the Lucas-Kanade method with pyramids. Then we analyze the performance of several filters for the preprocessing of the input images. Finally, we compare the results of the dominant plane extraction when using an arbitrary tolerance range and a dynamically computed one.

4.1   Optical Flow Parameters

We tested the Lucas-Kanade method with pyramids [3][6], using a number of different sizes in the integration window, as well as several pyramid levels. The results show that, with small integration windows, the images of the resultant dominant plane were too noisy, making the estimation of the free space hard or impossible to determine. In contrast, with large integration windows, we obtain better estimations of the dominant plane. As for the levels of the pyramid used to find the optical flow, results are better with pyramid levels of 0 and 1. This is expected, as bigger levels in the pyramid result in vectors of bigger magnitude. As a result, when computing the difference between the planar flow and the optical flow, the dominant plane presents bigger areas than those originally occupied by the obstacles in the image. A sample of these results is shown in figure 5.

We conclude that the best results for the dominant plane are those computed with an integration window of size 30 and pyramid levels of 0 and 1.

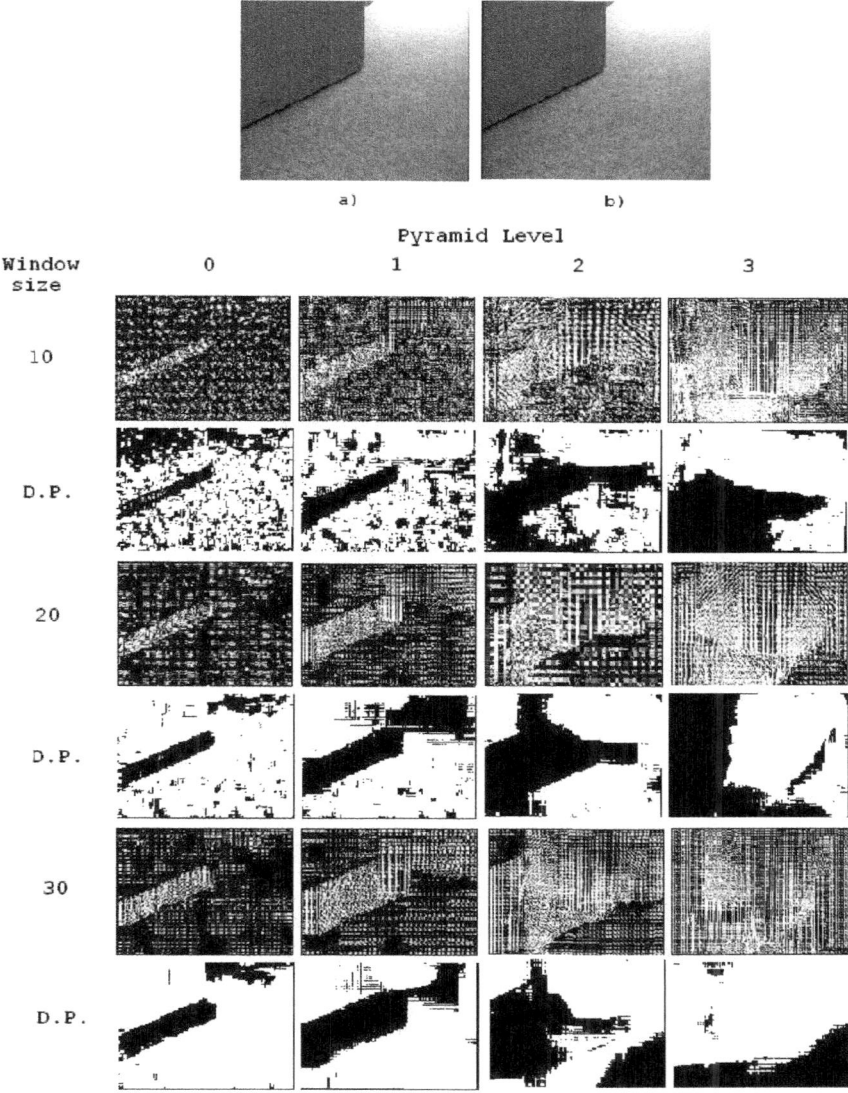

Fig. 5. The accuracy of the dominant plane of the images a) and b) is shown using several window sizes and pyramid levels. The accuracy increases with the size of the window and decreases when the pyramid level is bigger.

4.2   Filters

In order to reduce the noise in the images, we tested the performance of a gaussian filter and a bilateral filter. The bilateral filter tends to show better results when extracting the dominant plane and it better preserves the size and shape of the obstacles. The gaussian filter, on the other hand, generates images with a lot of noise if small kernels are used. If the size of the kernel is increased, the edges of the objects increase too, causing an increment in the area that the objects occupy in the dominant plane.

However, a less desirable effect when using the bilateral filter is a considerable increase in the processing time compared to the time taken when using a gaussian filter.

4.3   Dominant Plane Detection

The experiments from [10] show that estimating the dominant plane using a pre-defined value for the tolerance range is greatly affected by common fluctuations during the navigation. Such fluctuations include noise present in the acquired images, vibrations along the route, abrupt changes in the scene and illumination changes, among others.

These fluctuations lead to an increase in the time taken throughout the process and to a poor quality of the resultant dominant plane image.

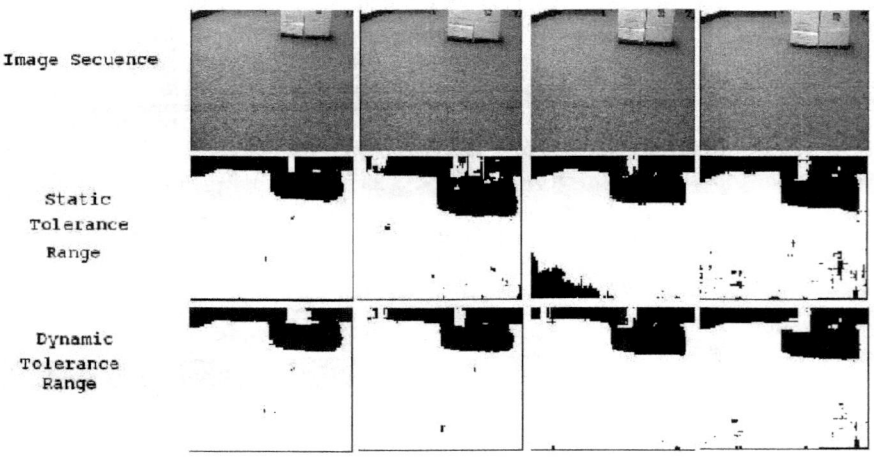

Fig. 6. Dominant plane of a sequence of images where the static tolerance range was set to 1 for the four images and values computed for the dynamic tolerance range were 1.24, 1.57, 1.27 and 1.54 for the dominant plane detection. In the figure, the dominant plane estimated with a dynamic tolerance range resulted in less noise in the dominant plane.

In all cases, problems with illumination need to be taken into account. For instance, reflections on surfaces of the floor or the presence of projected shadows that visually depend on the position of the illumination source. This is because the reflecting surfaces generate movement in the scene as if they were obstacles and these are seen when the optical flow is computed.

Estimating dynamically the tolerance range ε produces a robust and consistent dominant plane estimation especially during abrupt changes in the velocity of the displacement of the robot and when noisy images and vibrations in the camera are present. Additionally, the time taken during the extraction of the dominant plane is also drastically reduced.

Results of the estimation of dominant plane using a dynamic tolerance range in images of size 640x480 are shown in Figure 7. The figure shows the dominant planes estimated from a sequence of images. The values of the tolerance range vary between images and the dominant plane remains consistent in the sequence.

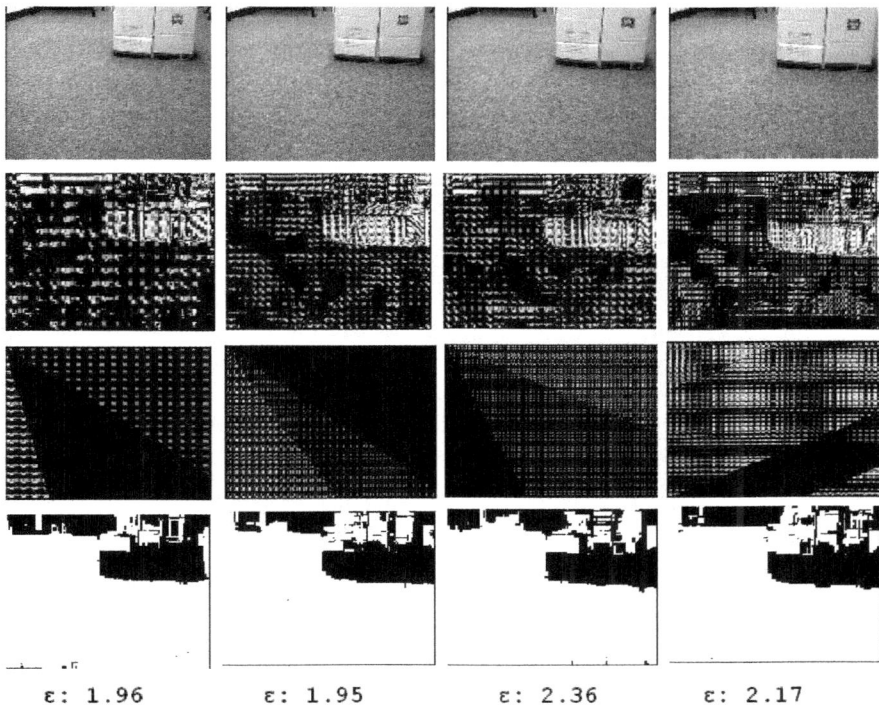

ε: 1.96 ε: 1.95 ε: 2.36 ε: 2.17

Fig. 7. The first row is the original image sequence. The second row is the optical flow of two successive images with an integration window of 30 and a pyramidal level of 0. The third row is the planar flow estimated from the affine coefficients and the fourth row is the dominant plane computed with dynamic values for the tolerance range calculated by the arithmetic mean using values selected from the vector field.

Because the displacement of a mobile robot is not always constant, the images can present abrupt changes that either could produce incorrect dominant planes or dominant planes that take too long to compute, given that most of the optical flow vectors are bigger than the tolerance range. The dynamic computation of the tolerance range greatly reduces this deficiency.

Our experiments were carried out using a sequence of images acquired with a Canon VC-C5 Pan-Tilt-Zoom camera mounted on a Pioneer 3-AT[1] mobile robot. The image sequence was taken at a rate of 20 frames per second, inclination angle of 21° and image size of 640x480, in an office corridor. All experiments were run on a 1.1 GHz Pentium M processor with 512 MB of RAM.

5   Conclusions

We have presented an improved method for extracting the dominant plane using a tolerance range computed dynamically in each pair of successive images. This method makes the estimation of the dominant plane adaptable to changes in the velocity of displacement present in the scene. The method is also more robust when processing noisy images and is less susceptible to camera vibrations. In terms of accuracy and efficiency, the performance of the dominant plane estimation is enough to successfully perform the navigation. We first used a bilateral filter with a consecutive pair of images to eliminate the possible noise. We then found the optical flow of the images and computed the value of the tolerance range ε using the arithmetic mean of the length of the optical flow vectors. Then we selected three random points to calculate the affine coefficients and compute the planar flow with them. The difference between the optical flow and the planar flow was computed to determine the points of the dominant plane. If the dominant plane occupied more than half of the image, the process was terminated. Otherwise, the affine coefficients would have to be computed again by selecting another set of random points. Our experimental results show that the parameters in the optical flow estimation are important in the detection of free space. If small values of pyramid levels and large integration windows are chosen, then the dominant plane can be calculated with a high degree of accuracy. The results presented in this paper show that the estimation of the dominant plane depends mostly on the optical flow field to produce appropriate values for the tolerance range. In addition, the use of a bilateral filter for removal of noise increases accuracy when extracting the dominant plane.

References

1. Barron, J.L., Fleet, D.J., Beauchemin, S.S.: Performance of optical flow techniques. Int. J. Comput. Vision 12(1), 43–77 (1994)
2. Beauchemin, S.S., Barron, J.L.: The computation of optical flow. ACM Comput. Surv. 27(3), 433–466 (1995)

---

[1] http://www.mobilerobots.com

3. Bouguet, J.-Y.: Pyramidal implementation of the lucas kanade feature tracker, description of the algorithm. Technical report, Intel Corporation, Microprocesor Research Labs (1999)
4. DeSouza, G., Kak, A.C.: Vision for mobile robot navigation: A survey. IEEE Transactions on Pattern Analysis and Machine Intelligence 24(2), 237–267 (2002)
5. Kim, D., Nevatia, R.: A method for recognition and localization of generic objects for indoor navigation. Image and Vision Computing 16, 729–743 (1994)
6. Lucas, B.D., Kanade, T.: An iterative image registration technique with an application to stereo vision. In: Proceedings of the 7th International Joint Conference on Artificial Intelligence (IJCAI 1981), pp. 674–679 (April 1981)
7. Mccarthy, C., Barnes, N.: Performance of optical flow techniques for indoor navigation with a mobile robot. In: Proceedings of the IEEE International Conference on Robotics & Automation, pp. 5093–5098 (2004)
8. Nakamura, T., Asada, M.: Motion sketch: Acquisition of visual motion guided behaviors. In: Proc. of IJCAI 1995, pp. 126–132. Morgan Kaufmann, San Francisco (1995)
9. Ohnishi, N., Imiya, A.: Featureless robot navigation using optical flow. Connection Science 17(1-2), 23–46 (2005)
10. Ohnishi, N., Imiya, A.: Dominant plane detection from optical flow for robot navigation. Pattern Recogn. Lett. 27(9), 1009–1021 (2006)
11. Santos-Victor, J., Sandini, G., Curotto, F., Garibaldi, S.: Divergent stereo in autonomous navigation: from bees to robots. Int. J. Comput. Vision 14(2), 159–177 (1995)
12. Vamossy, Z.: Pal-based environment mapping for mobile robot navigation. Technical report, Institute of Software Technology, John von Neumann Faculty of Informatics, Budapest Tech (2007)
13. Wolf, J., Burgard, W., Burkhardt, H.: Robust vision-based localization by combining an image retrieval system with monte carlo localization. IEEE Transactions on Robotics 21(2), 208–216 (2005)
14. Zucchelli, M., Santos-Victor, J., Christensen, H.I.: Multiple plane segmentation using optical flow. In: British Machine Vision Conf., 2002, pp. 313–322 (2002)

Do Cognition and Robotics Share Common Ground? Some Answers from the MCS Perspective and an RH-Y Case Study

Jean-Daniel Dessimoz

HESSO-Western Switzerland University of Applied Sciences,
HEIG-VD, School of Management and Engineering,
CH-1400 Yverdon-les-Bains, Switzerland
Jean-Daniel.Dessimoz@heig-vd.ch

Abstract. In the historical development of human tools and techniques, as well as of human philosophies, whereby the spiritual and material worlds tend to progressively merge, the moment has arrived where cognition seems also to have become a feature of man-made artifacts. The "MCS" - Model for Cognitive Sciences - provides a focused theory of core cognitive entities, including knowledge, expertise, learning, and intelligence. It is behavioral, is independent of implementation, and it provides a quantitative metric system. From the MCS perspective, cognition has fundamental limitations, which are inherited from properties of modeling and information. Cognition nevertheless has a potentially unique role in making both human-based and machine-based control effective. In the cognition domain, the boundaries between humans and robots depend on one's point of view: from the MCS perspective, in the core definitions and metrics, the overlap is total; when considering properties typically defined for a human context, such as consciousness, conscience, or life, a rather simple and direct analogous correspondence can also be given; now if an attempt is made to create robots that are identical to humans in all respects, this attempt can only fail, globally or in any significant part, and in particular for the cognitive subsystem. Examples taken from robots developed for Eurobot and Robocup at Home show that machine-based cognition is a common ground shared by robotics and cognition. Many instances show how the same robots prove to operate with peripheral cognitive properties classically defined for the human context, in particular, and newly here, the properties of consciousness, conscience and life. Nevertheless, many people intuitively feel that implemented robotics and human cognition can only remain totally disconnected; for any progress in this domain to become possible, it would be necessary to clarify which purpose is to be aimed at.

Keywords: Cognition, cooperating robot, Robocup-at-home, domestic help, cognitics, MCS theory, consciousness, conscience, life.

1 Introduction

Throughout the ages, humans have steadily succeeded in extending their basic natural abilities by the invention of new tools and techniques. In recent times, they have

D. Obdržálek and A. Gottscheber (Eds.): EUROBOT 2010, CCIS 156, pp. 14–29, 2011.
© Springer-Verlag Berlin Heidelberg 2011

dreamed of artificial servants, robots, as possible substitutes for themselves, especially in performing risky or uninteresting chores. Progress in robotics is now so advanced that robots and humans sometimes seem to be competing for leadership in a very controversial domain: cognition, i.e. the ability to "think", and more precisely, to process and/or generate relevant information.

Convergence of robots and humans in cognitive domain has been slow but is a fact.

Cognition has received the attention of philosophers for millennia. The Ancient Greek tradition (e.g., Pythagoras or Plato see, e.g., [1]) tended to split human beings into two parts, the first one more "spiritual", connected to ideas and gods, and the other one more "material", totally embedded in the surrounding physical world. In this context, cognitive aspects were evidently located in the former, ideal arena. The evolution of philosophy has progressively led to theories where these two domains were better and better integrated, to the point where their separation is impossible, as in the Constructivist theories of Jean Piaget [e.g., 2], or the notion of emergence in the systems of Varela and others [e.g., 3]. Cognition is no longer bounded to the world of ideas, nor even the brain, but is also seamlessly embedded in sensorimotor elements and even in the environment and the world at large.

Robotics is a young field of research. It is only in the last centuries that physical power has really been harnessed in machines, far beyond human and animal capabilities; industrial manufacturing has made available a variety of new structures and materials. These developments inspired the notion of robots, beginning about one century ago (re. K. Capek [4]). At that point in time, however, information was still a fuzzy concept, even though the telephone and the "recorder" had been invented, and there was as yet no idea of the possibilities to implement the more abstract, information-based, core operations essential in cognition. Then the computer was invented, improved, and miniaturized to the point, in the late 1990s, of being able to play chess better than the best human (re. Kasparov), and to be pervasive in human life.

Clearly, human related cognitive processes and the information processing capabilities of robots have already come very close together. It is the goal of this paper to delineate, in formal terms, what is the common ground between the two, and which elements remain distinctive.

The paper is organized as follows. Section 2 will review the main elements of the "MCS" model for cognitive sciences [5,7,8,14]. Section 3 will use the MCS perspective to underline the potential role and the essential limits of cognition. Section 4 will discuss the specific limits between machines and humans, in the cognitive domain, and further, it will introduce new definitions in the MCS ontology for the notions of consciousness, conscience, and life. Finally, Section 5 will consider detailed examples for the case of our RH-Y service robot, designed for domestic assistance, and its counterpart OP-Y, a mate with alternate locomotion properties.

2   Main Elements of the "MCS" Model for Cognitive Sciences

The MCS theory (Model for Cognitive Sciences) was developed as an effort to formally define the essential entities in cognitive domain, partly for a better understanding of cognition in general, but primarily in order to facilitate progress in the invention and design of novel, artificial (i.e., man-made) cognitive systems.

The MCS theory proposes a behavioral representation of cognitive agents. The most crucial items to consider are here the information flows incoming to and outgoing from cognitive agents; and time (see Fig. 1).

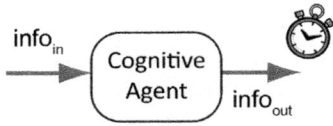

Fig. 1. Behavioral model of cognitive agents [e.g.5]

Information is essentially understood here, as proposed by Shannon, as being conveyed by messages, with a metric system based on probability calculus (re. [6] and Table 1); nevertheless, the current context is much broader than the initial technical communication context in which the theory was developed; for us, the information receiver is the more active and interesting cognitive agent to consider.

Table 1. Main concepts in the cognitive domain [e.g., 5]

Information:	$n = \sum p_i \log_2(1/p_i)$ [bit]
Knowledge:	$K = \log_2(n_{out} \, 2^{n_{in}} + 1)$ [lin]
Fluency:	$F = 1/\Delta t$ [s^{-1}]
Expertise:	$E = K \cdot F$ [lin/s]
Learning:	$\Delta E = E(t_1) - E(t_0); > 0$ [lin/s]
Experience:	$R = r(n_{in} + n_{out})$ [bit]
Intelligence:	$I = \Delta E / \Delta R$ [lin/s/bit]
relative Agility:	$Ar = \tau/T$

T: Fluency and communication delays
τ: Reaction time of target system, to be controlled

The most essential properties of cognitive agents are mainly associated with the notions of knowledge, i.e., the ability to deliver the right information, and fluency, which is related to the agent's processing time.

Taking a broader view, there appear other interesting ancillary or derived notions, such as experience, the integrated amount of incoming and outgoing information witnessed in the domain of interest, complexity, the amount of information necessary to describe a domain, or expertise, the product of knowledge and fluency.

A still broader view reveals other properties; in particular, further derivatives of high interest include learning, increase gained in expertise; and intelligence, the ability to learn from experience.

3 Potential Role and Essential Limits of Cognition from the MCS Perspective

The MCS theory sheds bright light on the serious cognitive limits humans and machines face. Nevertheless, when we also consider dynamics, especially in closed-loop configurations, the roles of cognition and its automated variant, cognitics, appear extremely relevant and often critical for success. Both points are further discussed below.

3.1 Limits of Humans and Machines in the Cognition Domain

The MCS theory provides necessary definitions and metric units for the cognition domain. It is, however, based on Shannon's information theory, and thus inherits serious limitations: the time variance and subjectivity of information, as well as, even more fundamentally, a severe necessity of modeling [7].

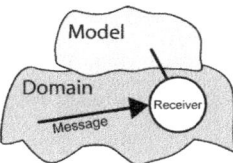

Fig. 2. Information is conveyed by messages, which allow cognitive agents (receiver) to form and update their opinion (model) relating to some subset of reality (domain)

Time variability. As shown in Fig. 2, the essential property of information is to update the receiver's model, and in particular to reduce a priori uncertainties. This means that p, the probability of occurrence of a possible incoming message, is de facto changed to 0 or 1 upon receipt of a message, depending on the nature of the message. Consequently, the amount of information conveyed by the incoming message immediately drops to zero.

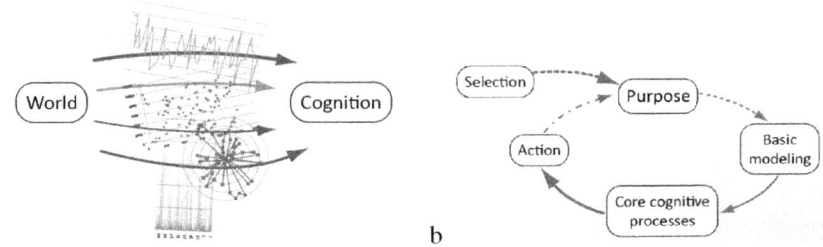

a   b

Fig. 3. Quantitative assessment shows that the complexity of the world is infinite, no matter how restricted the domain, and immediate cognition is consequently impossible (a). Nevertheless, when attention is focused on a specific purpose, it very often appears that crude, tractable, world representations (models) may be sufficient (b) and cognition may successfully proceed in this context.

Subjectivity. Shannon's fundamental equation for assessing information quantities involves p, the prior message occurrence probability, as perceived by the receiver: different receivers may have different models and a priori probabilities; thus, the information quantity may vary simultaneously for each of these.

Modeling. Let us adopt the same strategy for reality, for the world, as advocated for cognition in the MCS theory, namely, to adopt a quantitative approach. This usually leads to infinite amounts of complexity, even for very focused domains. Consequently, specific models must be devised that closely adhere to circumstances: hic et nunc (focus in space and time), ad hoc (for a specific purpose); see Fig. 3.

3.2 System Dynamics and Cognition

Cognition is impossible in a strictly static world, as cognitive agents always need to elaborate the right (outgoing) information. Now, although it is qualitatively necessary, time is usually not quantitatively critical; in such quasi-static views, cognition is already obviously useful, providing right (outgoing) information.

When dynamics are also considered, several new aspects appear, where cognition may also help in a variety of ways. We shall discuss below the case of control. This is done in two steps, in direct action and in closed-loop systems.

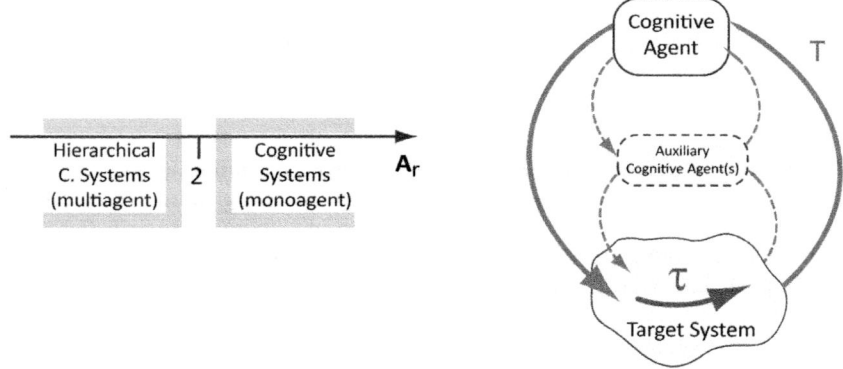

Fig. 4. An agent may successfully control a system in closed-loop mode only if its relative agility is better than 2. If the latter is smaller, a hierarchy of agents must be considered (Fig. from [9]).

Direct action. When a cognitive agent generates (outgoing) information and triggers action, dynamics show that controlled systems always require some time to react. If a result is now desired at a precise point in time, for example to get participants to reach the venue on the day of an international conference, a cognitive approach allows us to select the right moment in the future, and to anticipate all delays, launching all necessary operations at the optimal points in time to successfully reach the goal in time.

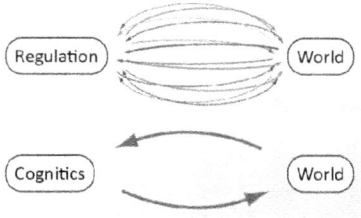

Fig. 5. Cognitics help in control by identifying relevant factors and processes, determining critical values, sparing measurements, reducing or sometimes even eliminating critical delays, anticipating effects and compensating for disturbances

Closed-loop control. Sometimes unknown factors, possibly random factors, disturb the state of controlled systems. To achieve success, the cognitive agent controlling the disturbed system must perceive the changes and adapt its actions. Schematically, two situations appear; in the first case, cognitive agents are agile enough and communication delays are short enough to allow for a solution involving a single agent; in the other case, control tasks must be distributed among several agents, each of them subject to the same constraint, in order to guarantee stability and to successfully reach their targets in their respective domain (re. Table 1, Fig. 4, and [8].).

In all cases, the situation may be improved as cognition allows forecasting phenomena and anticipating actions, thereby possibly compensating for potentially disturbing delays.

4   Specific Boundaries between Robots and Humans

In the previous sections, we have recalled the core properties of MCS, as well as the role and limits of cognition. The question in the title can now be addressed: do cognition and robotics share common ground? The answer is given below, first in the cognition domain, and then the scope of the question is expanded in two steps; to the fields of consciousness, conscience, and life, newly introduced in the MCS model; and finally to address the possibility of convergence of overall human and robotic natures.

4.1   Boundary in Cognition Domain

Considering the definitions of the MCS theory, our attention is strictly confined to cognitive abilities. In this context, the MCS model applies without distinctions to humans and machines, all such cognitive agents being represented as purely behavioral systems.

It is clear in this context that robots already have the possibility to do better than humans in quantitative terms: more bits of information perceived, more knowledge, more expertise, learning, intelligence, generated information, and so on.

Differences remain in various domains that relate specifically to those domains: physical perception (e.g., infrared is only perceived by some robots) and action channels (e.g., 10-fingered hands, with skin, are currently only the features of humans), language, exposure to news, repeatability, and so on.

4.2 Expanding MCS Theory for Robotics

The quest for robots for domains traditionally reserved to humans goes on. As a new contribution, the notions of consciousness, conscience, and life are discussed below, first in broader and somewhat informal terms, then with more focus and full compatibility with current MCS theory.

Consciousness. Consciousness is a property of cognitive systems, and can be defined to different degrees. The etymology of the word contains a root referring to cognition, to the ability of knowing, and a prefix referring to the subjective nature of this knowing. The very least degree of consciousness is simply awareness, the ability to know and thus to cognitively accompany what is going on in the world around the cognitive agent. The ability to react may be a sufficient indicator of consciousness according to this minimal definition.

A more demanding degree of consciousness calls for an additional, explicit, and regularly updated representation of what is going on around the agent.

A still higher degree of consciousness is attained when some aspects of the agent itself are explicitly present in the agent's representations; self-contemplation is performed. The scope of self-contemplation may vary, from some elementary self-aspects to more extensive ones, and even to the inclusion of "external" components, representing the environment in which the agent develops its activities.

In summary, for the MCS theory, consciousness is the ability of a system to perform its ordinary cognitive operations. Its value is essentially Boolean, corresponding to the presence or absence of consciousness. If required, the various degrees of consciousness discussed above should be addressed, in the MCS theory, as different, specific domains, for which the same definition of consciousness is applicable. In quantitative terms, this simple view can be complemented by a finer attribute, a *consciousness index*, defined as the ratio of the current level of operation to the ordinary level of operation. Here in practice an exclusive choice must be made, which is application dependent: either the Boolean view is sufficient, or the finer estimation approach is required instead.

Conscience. Conscience may be used as a synonym of consciousness, especially in the most demanding, self-oriented interpretation given above; in English, it includes a capability to judge the right or wrong character, in ethical terms, of an agent's decided actions.

Conscience implicitly requires not only that cognitive agents include in their cognitive domain representations of themselves and of their own behavior, but also that the agents include representations of their environment, with its associated operational modes. Only at this point can the possibility emerge for cognitive agents to compare their own behavior to the customary norms of the environment, and consequently, to estimate the right or wrong character of their operations (note that this description remains closely connected to the etymological meanings of the words ethics-environment and moral-customs).

For the case of our current robots, for laymen, ethics may not look relevant; it may seem hard to imagine that robots would ever attempt to break ethical laws. However, to make a right decision, the ethical question must first be addressed. Moreover, even in an environment of very moderate complexity, choices must be made in the face of conflicting ethical values, depending on the level of attention (e.g., a lower priority law is broken for the sake of a more general one). Of course, environments may be of various complexities, and may include, for example, groups of agents to which the agent with a conscience may relate in diverse ways. Notice that in the MCS models, as well as in reality, what is immediately apparent of agents is their behaviors. Consciousness, and its derived benefits, can still improve if agents express themselves and so share some of their internal representations; communication develops, e.g., flowers bloom, and animals develop a common culture.

In summary, for the MCS theory, conscience is the property of a cognitive agent whereby it includes in its cognitive domain some aspects of itself, of its own behavior, as well as of the environment and related customs; finally adapting its own actions as a consequence. The value that can be given to conscience is essentially Boolean, corresponding to the presence or absence of conscience. For finer quantization, all the core MCS notions essentially also apply here (e.g., information, complexity, knowledge and expertise); the possible specific differences in quantities relate to the respective, specific cognitive domains considered.

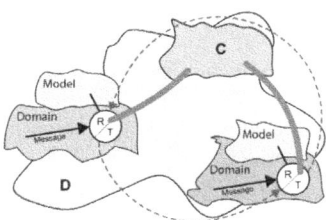

Fig. 6. Consciousness implies, in individual agents, some knowledge of themselves and of their environment. In a group, the degree of consciousness may be improved with appropriate communication among agents, and a common culture.

Life. Life is the property of agents to be able to perform their ordinary operations. Life can be defined in different grades of increasing requirements, which can also be viewed as related to the time duration of ordinary operation (functional activity), without discontinuity. Thus, a key unit seems here to be the time unit.

In its most basic form, life refers to the operational continuity of agents themselves.

A more demanding definition for life requires the ability of agents to actively sustain their own operations, and possibly recover from failures, thus possibly extending the duration of functional activity.

A still more demanding requirement refers to the ability to persist across generations: life thus allows agents to replicate themselves, having children capable of taking over ordinary operations for longer periods.

Time span of functionality can still increase, life being considered beyond the scale of a species, over evolutionary phases, and even, ultimately, at the scale of development of a whole life tree such as ours on planet Earth, from its very beginning, billions of years ago, to a yet undefined future.

From a practical perspective, the first basic definition given above for life is adequate, with variations in the possible requirements being equivalent to variant definitions for the cognitive agents under discussion (such as individuals, generations and species). A quantization may be useful in terms of life intensity, finer than just a Boolean value, namely life or death. A ratio (life intensity index) of current level of operation with respect to the ordinary level of operations might do it. This may be closely connected to several other notions in psychology (e.g., sleep, wakefulness, consciousness and arousal); this connections however are not really dealt with here, even though the notion of domain in MCS context would allow to do it rigorously. More generally the concept of life has led to many activities (e.g. [15], incl. an associated full page just for the purpose of disambiguation of meaning); no reference though could be found for specific questions addressing living versus non-living artifacts.

In summary, for MCS theory, life is the property of agents to be able to perform their ordinary operations. Its value is essentially Boolean, corresponding to life or death. In quantitative terms, this simple view can be complemented by two finer attributes, a *life intensity index*, defined as the ratio of current level of operation with respect to ordinary level of operation, and a *life time,* a duration measured in time units.

4.3 Should the Ultimate Robot Be a Human?

From the very beginning [4], there has been ambiguity in the definition of robots. Some observers have understood Rossum's creatures primarily as machines capable of providing flexible services to humans, while others have been concerned with the idea of artificially replicating humans.

Clearly, the approach of the first definition is easier to adopt, to develop robots according to requirements that are rather task-oriented, functional. Engineers typically favor this approach. It also offers additional benefits in terms of robustness and economy. Depending on the applications, the requirement of exact similarity between robots and humans is debatable: in the same way as a plane does not need moving wings like a bird in order to usefully transport people and goods, the practical solution chosen for a robot to fulfill specifications may often be validly different from human solutions; for example, a robot might rely on wheels rather than on feet. Yet as the number of specific duties transferred from humans to robots increases, there may be more advantages to approaches where robots are more akin to humans: humanoids (limbs more or less similar to humans) or androids (very similar look). Here some partial solutions are already welcome. Here also, if robots do their jobs better than humans, this is perceived as an advantage.

The second definition puts humans in the center of the scene. Robots should not only be capable of performing duties similar to those of humans, but moreover, robots should ideally proceed in the same way; ultimately, robots should be human in all respects. Obviously, this task is impossible to fulfill by non-biological means. Also in biological terms, if required, the most promising road would theoretically be to improve cloning techniques, and then to concentrate on education, but such a road is of course completely ethically unacceptable.

To advocate the first approach again, it is worth noting that, in the second case, humans are already not only the result of a well-defined, linear genetic legacy: in biology, some genes are acquired by lateral transfer, and some other genes are often randomly changed by external influences; moreover, when living, humans are more and more complemented with artificial accessories: spectacles, auditory aids, pacemakers, wooden legs, artificial forearms and hands, infrared goggles or telescopes, cars, cellular phones, tools, pharmaceuticals and drugs, and so on. Bionics, cyborgs and avatars are examples of engineered extensions of the human body. This approach may have some value in understanding better humans, but let us state here that it is *not* our goal to develop robots just as replicas of humans. Even if it were the case, we would have to face the fact that we are very far from reaching such a goal. No robot is in sight yet, so akin to humans that a risk of schizophrenia would threaten it: am I a human or a machine?

5   The Case of RH-Y, OP-Y and Other Robots

Section 5 will consider detailed examples: of our RH-Y service robot, designed for domestic assistance, and its fellow OP-Y, a counterpart with alternate locomotion properties. The plan is here the same as above, to address, in sequence, the boundaries between human and robotic cognition, as well as the human-like properties newly defined above in the MCS framework: consciousness, conscience, and life.

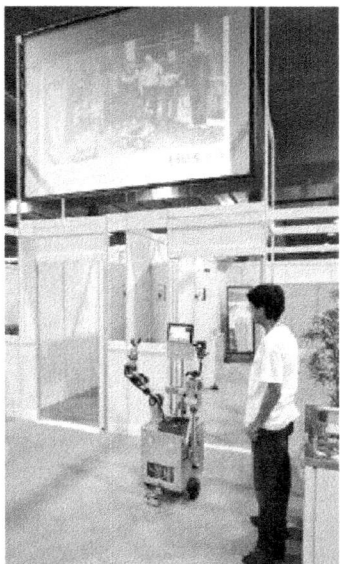

Fig. 7. General view of RH4-Y robot in Graz during the "Introduce" test, where the robot autonomously enters home, moves to the living room, talks, presenting itself and team, then leaves

5.1 Do Cognition and Robotics Share Common Ground in the Case of RH-Y?

Consider RH-Y in the Robocup context, for example, as illustrated in the "Introduce" task (Fig. 7), as well as in the "Follow me" and "Order by gestures" tasks (Fig. 8). It will be shown that, in some sense, some tasks usually performed by humans and considered in the past as requiring cognitive capabilities can be taken over by robots; but it will also be shown that some gaps remain that some observers will consider as requiring cognitive abilities, that possibly in their views relate exclusively to human nature.

Cognition is an important capability of robots such as RH-Y. In the "Introduce" task, for example, RH-Y can speak, move in space, and adapt its elementary actions in order to compensate for disturbances and to progressively reach appropriate locations.

There are of course always limits on what can a system do. However, the argument here is that, in principle, the solution usually lies adding quantitative precision, and not necessarily in qualitatively adding new concepts, thus elaborating models of increasing complexity. For example, for a human attempting to jump over a wall, the height of the wall is critical for success, and it is therefore appropriate to take the height measurement into account.

RH-Y can virtually speak thousands of words in "English" and move with accuracy on the order of centimeters at speeds up to about 1 meter per second on flat ground. All this translates into specific quantities of information, knowledge, expertise, and so on. For example, one word among 10,000 (equiprobable) words may be estimated as conveying 13 bits of (word) information. Considering the elocution of a single word at a time, 13 bit are present at the speech synthesizer input side, and $2*20$ kHz$*7$ bits of (sound wave) information per second are generated on the output side; considering that words are uttered at a rate of one per half a second on average, the amount of knowledge is of about $13+19=32$ lin of knowledge; and this yields an expertise of 64 lin/s.

Fig. 8. Having entered home by following a human guide, OP-Y, on the right, moves according to gestures performed by its counterpart, the RH4-Y robot, in Graz, July 2010

Or is cognition purely human? Unfortunately, some observers consider cognition to be a capability exclusively associated with human nature; thus, they will consider that since RH-Y speech synthesis can be performed by (now very common) electronic and microtechnological means, this is de facto proof that the speech act requires no cognition. Moreover, as RH-Y features loudspeakers rather than vocal chords, a tongue and lips, it is even more obvious that such robots are just machines, and by no means human; actually for those observers, even the name of robot could be disputed here; "machine" would be more fitting!

Synthesis. The two above paragraphs give two opposite views, the first one, compatible with the MCS model, confirms that machine-based cognition is possible, and just accounts for one of many abilities of robots, including perception, action, locomotion and communication. Cognition is here totally integrated in robotics. Nevertheless, it can be accepted that cognition also develops in domains other than robotics, thus the question in the title can be answered affirmatively: there is a common ground shared by robotics and cognition. The second view appears intuitive, emotional, and scientifically unfounded from a behavioral perspective; nevertheless, many people feel that way, and in such a context, implemented robotics and human cognition can only remain totally disconnected; for progress to be possible here, it would be necessary to clarify the concepts involved and, first of all, the goal being attempted.

5.2 Illustration of New Human-Like Properties in the Case of RH-Y

Even though the direct requirement for notions such as consciousness, conscience or life remains debatable, the following arguments do have some value for the good functioning of machines and robots: the potential for better human-robot communication, the legacy of millennia of cultural developments in the human context, and finally also a better understanding of human nature.

Consciousness. What degree of consciousness characterizes RH-Y? 1. In the least demanding sense proposed above, awareness, consciousness is implicitly proven by its proper, ordinary operation, which, in many ways, involves adequate instantaneous reactions (e.g., arm joint motor torques) to external stimuli (e.g., wheel encoders, etc.). 2. At a higher degree of consciousness, awareness extends to what happens around the agent; this is done in several ways in RH-Y (e.g., microphone input, laser sensors, camera signals, and so on). 3. The next degree of consciousness involves, for RH-Y, the ability to explicitly represent internal and peripheral elements. Fig. 9 displays many examples of such capabilities. RH-Y has such an advanced state of explicit representations, both of internal elements and of external elements, that most of its operations can even be performed in simulation.

In quantitative terms, RH-Y in ordinary operative cognition can be said to have consciousness. The consciousness index introduced above could usefully refer to the completeness of its operational capability at a given point in time, e.g., in noting what percentage of its knowledge is available as a consequence of mounted, active

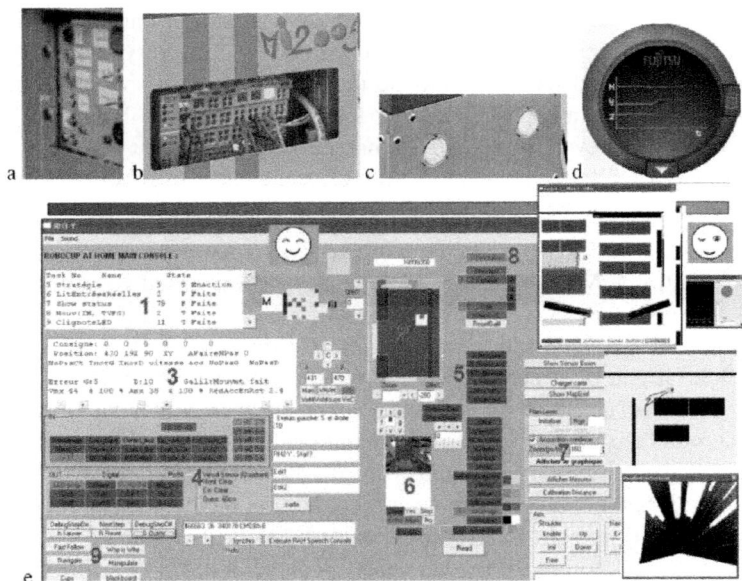

Fig. 9. The consciousness of our robots [12] is made evident by the expression of status and intentions as classically described by panels, displays and screens (a: Hornuss, b: Dude, both developed for Eurobot, c: programmable "eyes" or modulated headlights, RH3-Y; d: 3-D acceleration components on supervising computer; e: set of interactive control screens for RH3-Y).

perception, cognition, and action channels. The latter may greatly vary in time, and typically match specific tasks being benchmarked: such as autonomous navigation, 3D TOF ranger-based SLAM and human face recognition.

Conscience. As a synonym for consciousness, in its more demanding sense, conscience is demonstrated above for the case of RH-Y. Furthermore, we can consider, as an example of choice between two conflicting goals, the case of the "Follow me" task, as defined in Robocup at Home competition [10]. At a particular checkpoint, robots are approached by two humans, and a decision must be made to follow the "right one". The latter should be recognized as the guide introduced in a preceding phase. Another interesting example of the change of perspective of robots, with consciousness and self-reflection, is the sequence shown in [11], where robots gradually learn how range sensor data are best sequenced, with an egocentric perspective and an assumption of stability and continuity of its environment, subsequently taking a next step, which leads to the persistence and stability of the environment, and in this context, to the explicit representation of the self, as a mobile and situated agent!

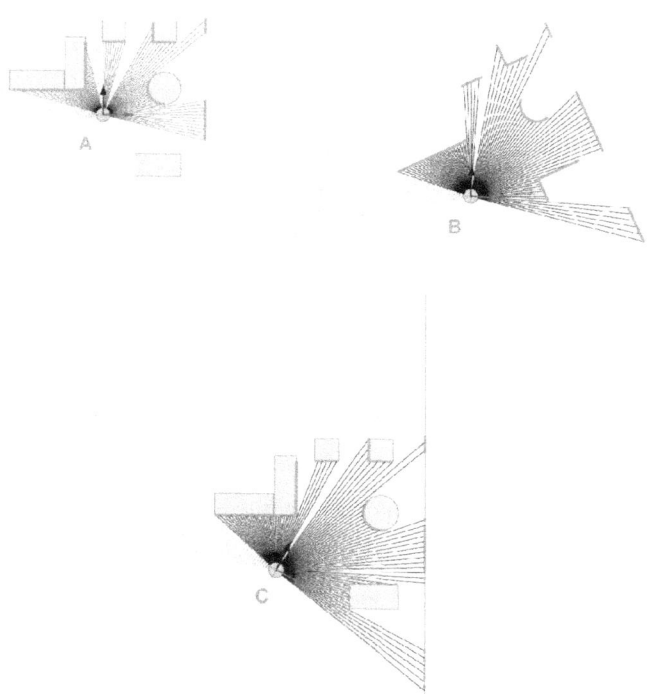

Fig. 10. An evident case of consciousness in our robots is illustrated here, where the application might relate to SLAM, pattern recognition, or positional calibration. In A, real egocentric distances are perceived with a scanning laser range finder, world elements are represented as green elements and are learned as a persistent model of the world. Later on, in B, the robot is in another location, with unknown orientation, along with the corresponding range data. A correlative process then matches data in B with elements of A. Consequently, the robot knows its own location in the world, C.

Life. RH-Y robots live only according to the most basic definition: they are functional when appropriately prepared and rely on careful human-based maintenance for sustained operation. An interesting feature of our robots, continuously present since their first edition in Eurobot context (Diego[3], 1998), has been to display a blinking signal, normally expressing correct operational status. For observers looking for analogies with humans, this can be considered in a way similar to a heartbeat, in order to facilitate checking whether the system is alive. For our robots, when operational, the life intensity index should typically be equal to one. An interesting exception is the case of our "Arthur" robot developed for Eurobot context (1999), which had a selectable "turbo" mode, by which the current limits of the motors could be switched to higher values, yielding larger instantaneous torques but thereby also reducing the expected lifetime.

Analogies with the human case could be carried further. For example, as relating to sustainability, in the same way as humans often need society and hospitals for some care, machines typically require maintenance. It is not clear, however, how extrapolating further analogies with the biological world might be useful or inspiring (e.g., calling for robot hospitals).

6 Conclusion

In the historical development of human tools and techniques, and more recently of machines and robots, as well as of human philosophies, whereby the spiritual and material worlds have tended to progressively merge, the moment has arrived where also cognition seems to have become a feature of man-made artifacts.

The "MCS" model for cognitive science provides a focused theory for core cognitive entities, including knowledge, expertise, learning, and intelligence. It is behavioral, is independent of physical implementation nature, and it provides a quantitative metric system based on information and time.

From the MCS perspective, cognition has fundamental limitations related to the domain where it develops: modeling is necessary and yet always infinitesimally complete; information has a very short decay time and is subjective. Cognition may nevertheless have a potentially unique role in making both human-based and machine-based control effective, with single or multi-agent architectures, even when large delays are necessary for action, or circumstances change highly dynamically.

In the cognition domain, the boundaries between humans and robots appear very differently, depending on one's point of view: from the MCS perspective, no difference whatsoever is made in the core definitions and metrics between humans and robots, so the overlap is total; when the perspective widens, considering properties typically defined for human context, such as consciousness, conscience, or life, a rather simple and direct analogous correspondence can also be given. Now on the contrary if, beyond full similarity in *behavioral* modeling and *functional* operation, a further attempt is made to create robots *identical* to humans in all respects, this attempt can only fail, both globally and for any significant part, such as for legs, eyes, skin, growing paradigms, self-sustenance and repair abilities; in particular here, the attempt to create a cognitive subsystem identical in all respects to the one of humans, i.e. much beyond the equality of (blackbox-type) behavior, and functionality, this attempt can only fail.

Examples taken from robots developed for Eurobot and Robocup at Home, e.g., speech synthesis, show that machine-based cognition is a common ground shared by robotics and cognition. Many instances show how these robots prove to operate with peripheral cognitive properties that were classically defined for human context, in particular consciousness, conscience and life. Nevertheless, the fact remains that many people feel on a visceral level that implemented robotics and human cognition can only remain totally disconnected; for any progress to be possible here, one must clarify the purposes and goals of such progress.

The author wishes to thank the anonymous reviewers for several helpful suggestions.

References

1. Ross, W.D.: Plato's Theory of Ideas, p. 250. Clarendon. Press, Oxford (1951)
2. Piaget, J.: The Child's Construction of Reality. Routledge and Kegan Paul, London (1955)
3. Varela, F., Maturana, H.: Autopoiesis and Cognition: The Realization of the Living. Reidel, Boston (1980)
4. Karel Capek, R.U.R (Rossum's Universal Robots) (Rossumovi univerzální roboti) (1920)
5. Dessimoz, J.-D., Gauthey, P.-F.: Quantitative Cognitics and Agility Requirements in the Design of Cooperating Robots. In: Gottscheber, A., Enderle, S., Obdrzalek, D. (eds.) EUROBOT 2008. CCIS, vol. 33, pp. 156–167. Springer, Heidelberg (2009)
6. Shannon, C.E.: A mathematical theory of communication. Bell System Technical Journal 27, 379–423, 623-656 (1948)
7. Dessimoz, J.-D.: Contributions to Standards and Common Platforms in Robotics; Prerequisites for Quantitative Cognitics. In: International Conference on Simulation, Modeling, and Programming for Autonomous Robots (SIMPAR) 2008. First International Workshop on Standards and Common Platform for Robotics, Venice, Italy, October 3-7 (2008)
8. Dessimoz, J.-D.: Cognition Dynamics; Time and Change Aspects in Quantitative Cognitics. In: Second International Conference on Intelligent Robotics and Applications, Singapore, December 16 - 18 (2009)
9. Dessimoz, J.-D.: Cognition for a Purpose - Cognitics for Control. In: 4th International Conference on Cognitive Systems, CogSys 2010, ETH Zurich, Switzerland, January 27th & 28th (2010)
10. RobocupAtHome League (2010), http://www.robocupathome.org
11. Kuipers, B.: How Can a Robot Learn the Foundations of Knowledge? In: 4th International Conference on Cognitive Systems, CogSys 2010, ETH Zurich, Switzerland, January 27th & 28th (2010)
12. Dessimoz, J.-D., Gauthey, P.-F.: RH4-Y – Toward A Cooperating Robot for Home Applications. In: Robocup-at-Home League, Proceedings Robocup 2009 Symposium and World Competition, Graz, Austria, June- July (2009). Also: West Switzerland Univ. of Applied Sciences, Jean-Daniel Dessimoz: website for HEIG-VD RH4-Y cooperating robot for Robocup-at-Home context (March 2009), http://rahe.populus.org/rub/3
13. Dessimoz, J.-D.: La Cognitique - Définitions et métrique pour les sciences cognitives et la cognition automatisée. In: Roboptics (eds.), Cheseaux-Noréaz, Switzerland (August 2008), http://www.lulu.com ISBN 978-2-9700629-0-5
14. http://en.wikipedia.org/wiki/Life (April 25, 2010)

Elements of Hybrid Control in Autonomous Systems and Cognitics

Jean-Daniel Dessimoz

HESSO-Western Switzerland University of Applied Sciences,
HEIG-VD, School of Management and Engineering,
CH-1400 Yverdon-les-Bains, Switzerland
Jean-Daniel.Dessimoz@heig-vd.ch

Abstract. Intelligent systems and robots have made significant progress in recent years, in artificial intelligence (AI), situated automata, reactive systems, classical control, computer infrastructure and networks. The paper shows that hybrid approaches (i.e. a mix of these mentioned techniques), and in general cognitics (i.e. automated cognition), offer multiple benefits: assessing various cognitive elements, predicting phenomena, compensating for disturbances, embedding programmed systems in reality, estimating entities in virtually all regions of high dimensional spaces, learning and being expert, and running on digital processors with the best software methods. Additionally, some concepts are newly discussed, in order to facilitate a coherent global view: deliberation, top-down approaches, creativity and ingenuity are also important items in the general picture. The paper finally also addresses learning; it appears that only chance has the cognitive power to yield truly novel models, while expert resources remain necessary to collect, optimize, and make use of them effectively.

Keywords: Autonomous systems, Robots, AI, Control, Cognitics.

1 Introduction

Intelligent autonomous systems, and robots in particular, are continuously improving their ability to take the right steps towards mission goals that humans assign to them.

Control for such systems, including higher-level layers, must be endowed with algorithms or schemes that have not fully stabilized yet and also deserve special attention to the possibility of hybridization of the best current theories and practices (e.g. [4]). Four components have proven particularly crucial in the evolution of intelligent and automated systems, which we briefly review here: AI, situated automata, closed-loop and networks; this is not exhaustive as the trend toward more complex domains calls for a growing number of approaches, which cannot be all referenced here. The strategy of hybrid control is to integrate several of these components, as none of them seem capable to solve all relevant issues on their own.

AI. One of the four crucial components just mentioned is AI, as explored in a variety of ways, and it has reached such dramatic achievements as proving some

D. Obdržálek and A. Gottscheber (Eds.): EUROBOT 2010, CCIS 156, pp. 30–45, 2011.
© Springer-Verlag Berlin Heidelberg 2011

mathematical theorems or playing chess beyond the limits of most talented humans. In this regard, the most universally recognized definition of intelligence in the field today is that of Alan Turing, according to which the information-based [6], cognitive behavior of machines should be indistinguishable from that of humans [7].

Situated automata. Rodney Brooks made the revolutionary suggestion that modeling and world representations could be avoided, that machines could immediately perceive the world, and react; the key idea is to imbed the intelligent agent in reality as a situated automaton. More elaborate concepts have led to the notion of subsumption architecture [1].

Closed-loop control. For more than 50 years, control based on feedback measurements, has gained a whole corpus of contributions, which today allow, for example, for the fast and accurate positioning of minute magnetic heads on dense hard disks as well as for the successful travel of space rockets towards precise targets at astronomical distances.

Computer infrastructure and networks. A fourth major component in the successful evolution of intelligent and automated systems is provided by computer sciences and engineering, as well as by novel 1-1 and network-based communication possibilities.

However, significant limitations remain. For AI, some researchers (e.g., [1]) have pointed to the huge difficulty of representing, in general, objects and processes. Quantitative assessments also confirm that this is especially true when the goal is to encapsulate the essence of reality.

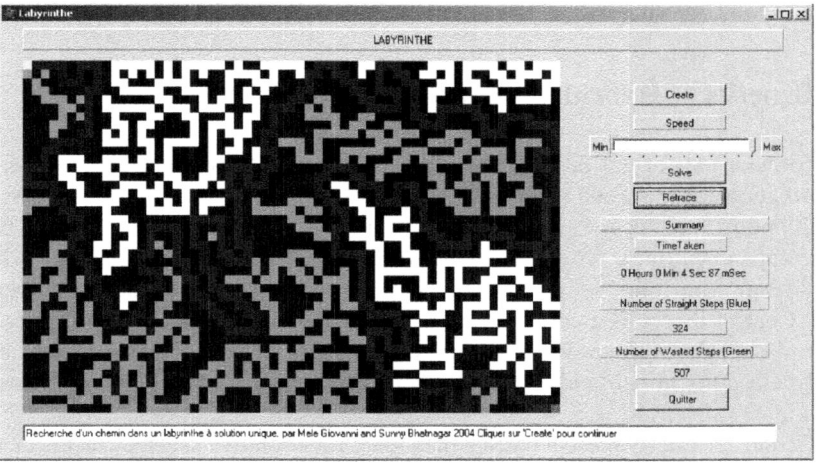

Fig. 1. Navigation in random mazes (walls are black, path is initially white) by a reactive mobile system without memory: starting in the upper left corner, the autonomous system maintains contact with the right "wall," thus successfully reaching the exit. Its path includes both the optimal path (displayed in blue) and about half of the dead ends (displayed in green). Other navigation examples with situated automata developed in our lab include stochastic strategies, for other, possibly dynamically changing contexts.

When one wants to avoid modeling and to connect directly to the world, one is constrained to "here and now" implementation; no past nor future world can be perceived so directly, neither alternate nor remote locations can be sensed, to say nothing of hypothetical, virtual worlds. Those strong constraints limit the possible benefits of purely reactive systems.

Real-world systems always suffer from delays, which are crucial for closed-loop systems behavior and performance (stability, speed, accuracy, and cost.).

This paper aims to contribute to pushing those limitations farther away. It proceeds in three main ways, presented in the three main parts of the paper . Part 2 shows how a hybrid approach can help, and how this is particularly true from the perspective of the Model for Cognitive Sciences,(MSC), which is central to cognitics, i.e., to automated cognition. Part 3 allows for the inclusion of new terms, in particular overlapping with the RAS-HYCAS[1] initiative, into the MSC ontology. Finally, Part 4 presents a broader overview of generic schemes for goal-oriented cognition and efficient control. In synergy with these themes, six figures illustrate the main concepts, often including hybrid views: 1. a reactive system and quantitative environment for cognition, 2. a schema of information based behavioral modeling and pyramid of core cognitive properties in MSC, 3. an automation test-bed, for benchmarking classical and AI techniques (fuzzy, neuronal, dimensioned with GA, multimodal), 4. an extension of classical control capabilities by a fuzzy control paradigm, 5. relative control agility as a critical and universal indicator for appropriate structures in (possibly complex, multiple, nested) closed control loops, and 6. RH4-Y, a cooperating robot, shown in world-class competition, featuring a balanced and effective original system that integrates numerous advanced technical and cognitive capabilities.

2   Benefits of Hybrid Approach and Cognitics

A particular idea of this paper on hybrid approaches is considering how to synthesize contributions from AI and reactive systems, the latter being understood in the sense of subsumption architectures. Successful synthesis is indeed possible, which is even more apparent from a "cognitic" (automated, cognitive) point of view and with due consideration for other classical fields: control and computer sciences and technologies. Control theory is a comprehensive framework where reactive systems can find their place, with well known advantages and limits. Computer technologies and networks are other broad fields featuring their own distinct and well known advantages and limits.

Section 2.1 briefly presents MSC, the model for cognitive sciences that is central to cognitics. This is a good preparation for sketching, in Section 2.2, the expected benefits of a hybrid approach – including not only AI and reactive behavior, but also the other fields just mentioned. Section 2.3, presenting considerations typical of a quantitative cognitics approach, opens the circle of possible contributors to hybrid solutions even wider, to include chance and time.

---

[1] Robotics Autonomous Systems – Hybrid Control of Automated Systems [4].

2.1 Quantitative Cognitics

The field of quantitative cognitics and the MSC model for the cognitive sciences [e.g., 2] have already proven useful in precisely defining the main properties of cognitive systems (knowledge, complexity, expertise, learning, etc.), in quantitatively assessing them, and in drawing conclusions from our experiences with them. Here are some examples that can be rigorously addressed in this framework:

1. The quantity of intelligence (unit : lin/s/bit) that characterizes a specific cognitive agent.
2. The ratio of complexity between a given domain of reality and a corresponding model of it.
3. The fact that there are many intelligent man-made artifacts; in particular current ordinary computers, capable of learning, e.g., with the cache paradigm.

The definitions provided in the MSC ontology are not claimed to be true (they describe an infinitesimal part of reality), but just useful; in particular, they are useful in engineering, in autonomous systems, providing the ability to take the right steps towards mission goals.

The cognitics perspective provides a sound basis for assessing and complementing classical AI and reactive architectures, as well as control, computer and software bases.

Now we seize the opportunity to raise interest in the perspective offered by MSC. Furthermore, in Part 3, new terms will be added to the ontology.

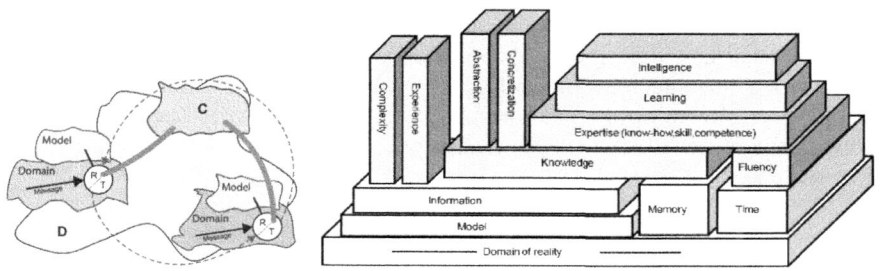

Fig. 2. In the MSC framework [e.g., 2], autonomous systems can be represented as behavioral systems (e.g., individuals or groups, on the left), and on the basis of information and time, their main cognitive entities can be quantitatively estimated (green concepts, on the right)

The MSC model is built on information theory and therefore inherits both the benefits and the essential limitations of Shannon's concept of information. In addition, the quantitative approach makes it clear that no domain of reality can be described by a finite amount of information. Consequently, a "Copernican" revolution is necessary in our approach: do not start from reality and deduce consequences, but start instead from the goal, freely selected. Then, hopefully, it will be possible to create/retain a finite model that minimally represents only those aspects of reality that are useful for reaching that particular goal.

2.2 Hybrid Approaches

Here we consider a variety of hybrid domains, various mixes of the above mentioned approaches: AI-Cognitics, Control-AI, Reactive systems (RS)-Cognitics, and AI-Computer technologies. Improvements have been noticed and can be similarly expected in numerous applications.

AI-cognitics. AI has so far been defined in a variety of ways that have not contributed much to its development. The most common definitions include the following:

A- Turing's definition, which is excessively anthropocentric and culturally oriented,

B- Social and epistemic definitions, which integrate what(ever) is done in self-identified AI labs through the world,

C- most commonly, the implicit definition, which is "suicidal" to the field of AI, that finds intelligence to be "that very cognitive property that specifically discriminates humans from machines,"

D- in the MSC model, intelligence is defined there as the specific property of cognitive systems that are capable of learning (learning and other features are given precise definitions, along with measuring units, all ultimately based on information and possibly time).

Except for case C, the definitions provided in the MSC ontology for cognitics are not necessarily incompatible with the first definitions above. More importantly, the benefit to be expected from adopting MSC in AI is the rigorous and encouraging assessment of numerous existing intelligent systems, as well as the possible quantification of other cognitive properties and requirements. A hybrid view is particularly natural here: in as much as the field of cognitics corresponds to automated cognition, it includes AI. Reciprocally, in classical terms, the field of AI is usually understood in a very broad sense, which includes all aspects of cognition, and not specifically intelligence alone.

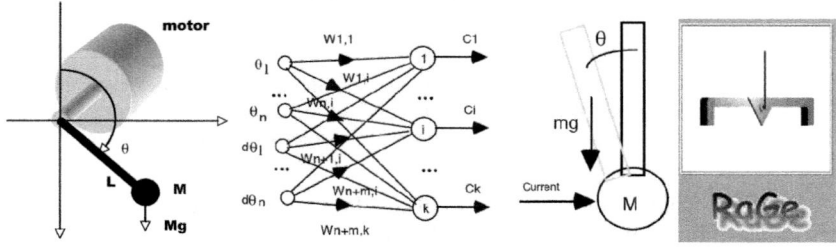

Fig. 3. Many physical processes behave schematically like the motorized inverted pendulum (left) and can be successfully controlled, using classical approaches as well as typical AI systems (in particular, fuzzy logic and neural networks) or mixed type systems, such as controllers synthesized with genetic algorithm paradigms (HEIG-VD examples)

Control-AI. As will be elaborated below, control processes sometimes need to rely on some feedback components, and the effective management of delays is critical for the effective operation of such closed-loop control systems. For years and years, it has been successfully verified that it is often possible to compensate for some delays, typically in practice by a prediction strategy based on signal derivatives. Now in hybrid terms, many new possibilities arise when AI is used in control systems as a mean to improve predictions, thereby to improve effectiveness, and even, in many cases, to automate novel applications, such as for robots cooperating with humans in cognitively demanding applications.

RS-Cognitics. Agents in MSC theory are the same as in reactive systems or situated automata, in the sense that they all involve essentially behavioral elementary components, i.e., they can be completely described by the input-output associations they feature (and time). A hybrid strategy may thus be useful, as RS will keep things simple where possible, while approaches advocated in cognitics can address factors beyond just the immediate signals emanating from the environment (e.g., contextual information and remote location modeling) as well as beyond just the strictly present time (past, future, simulations and "if-worlds").

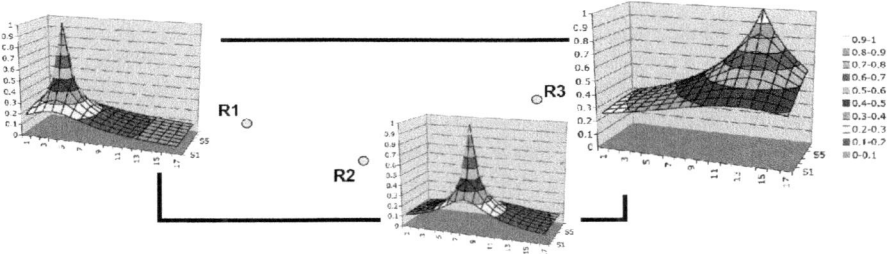

Fig. 4. An example in which a hybrid approach merges the control power of multiple classical controllers (R1, R2, R3), each individually and optimally tuned in key configurations, with the paradigm typical of fuzzy logic controllers (re. defuzzification), by which dynamic weights and interpolation functions (re. membership functions) ensure the smooth transitions in physical space (here, bidimensional) of the respective controller contributions

AI-Computer technologies. Let us consider the mix of AI (or more generally cognition) and computer technologies. In one sense, computer technologies obviously bring a lot to AI: practically all cognitive systems today are implemented on computers (and microprocessors), even if sometimes, conceptually, other implementation media are discussed (e.g., simulation of neurons, generations of evolving populations, virtual ants, etc.). Progress in microelectronics, software technologies, communication and network engineering is such that the processing of large amounts of complexity, knowledge, expertise, learning (to name a few core cognitive properties) is nowadays routine, yet a quantitative approach quickly proves that cognitive processes can yet be much more powerful than such tools and techniques! A striking example is provided by the accurate estimation of the number "Pi," i.e., 3.14… While boosting computer technology by a factor of 10 billion on speed (i.e., far beyond all current expectations and hopes), using a basic series for estimating Pi, would yield a mere additional *10 digits* compared with computing at current speed, (human) ingenuity has added about *200 millions digits* of accuracy to estimates of Pi in the past two decades! (re. App. A).

2.3 Ingenuity, Chance and/or Time

Cognition is necessary but probably not sufficient for strong innovation. Cognition is necessary for managing complexity, knowledge, expertise, abstraction, concretization and many other information-based entities and processes. Today, a huge number of (artificial) cognitive systems routinely run successful operations and deliver information that is physically impossible to be stored a priori, in particular, impossible to be integrally collected from experiments or to be precompiled even with human help[2].

But how can we create new models and novel cognitive systems? How can we make quantum leaps in improvements (such as the above case of estimating pi)? It is tempting to think of yet another cognitive property – ingenuity. However, any attempt to quantify this property meets serious problems. Either we must conclude that so far the basis on which to estimate quantitatively this cognitive property is still lacking, or we could view ingenuity essentially as just a regular cognitive process, here immediately embedded in a specific domain of *reality*, which would call for infinite amounts of knowledge and expertise.

In fact, there *is* a powerful possible source for strong innovation: chance.

In the MSC model, the theory shows that random processes, as they are capable of generating an infinite amount of unpredictable information, de facto prove to feature an infinite amount of knowledge in their domain.

Thus, counter-intuitively, we could consider ingenuity not as a specific cognitive property, but rather as a regular cognitive process that evaluates and keeps selecting the best of several (often of many) models or cognitive systems that are randomly generated by external, random sources, i.e., chance. This strangely brings us back to one of the most fundamental paradigms in AI: trial and error.

Chance is a possible source for innovation, but it may take a lot of time for success; here cognition (experts) may bring advantages by keeping focus on minimal sized domains to explore, keeping track of improvements, and by possibly tuning-up contingent solutions.

3 Defining New Terms in Cognitive Ontology Context

Hybrid approaches, such as the recent HYCAS initiative [4], bring attention to multiple interesting concepts. Some of them, such as learning, are central for cognitics and have already been listed and defined in the MSC ontology; some other ones have not, or not yet, such as control, reactive control, deliberation, top-down and bottom-up approaches. And other entities, including those mentioned above, including ingenuity and chance, or still others, such as creativity should all be discussed in the MSC

---

[2] In quantitative terms, a rough and conservative upper bound on the knowledge K, for which a direct memory-based implementation may technically be possible, can be estimated at 1000 lin.

framework or from the MSC perspective in order to have a more coherent and complete view; they are therefore addressed below, in the same order as just introduced. As an exception to this rule, creativity will be addressed before ingenuity.

Control. Generally speaking, control is a process that delivers commands to a system so as to reach some specific goal. While in general the words "control" and "commands" imply power, forces and/or other physical entities, here only the information aspects are considered. Thus, in the MSC model a control unit or agent is just a regular cognitive system, which, like any other one, is fully described by its input – output information flows ("behavior") and time; all derived cognitive entities (knowledge, expertise, complexity, etc.) are equally applicable here.

Reactive control and other control types. Reactive control is a particular type of control in which some particular kind of input information is relevant. A still more specific type of control, a subset of reactive control, is called closed-loop control, in which some input information entering the control system results more or less directly from the commands (i.e., output information flowing out of the same control unit) issued to the system being controlled (this specific input information is called "feedback"). The opposite of reactive control would be proactive control, in which information is autonomously generated by the control unit and is transmitted as "feedforward", "open-loop" information toward the system being controlled. When integrated in complex systems, control units may simultaneously feature multiple control types, depending on the subsystems and functions considered.

Deliberation. The root of the word deliberation refers to Roman scales by which weights were measured. The basic test, "which side of the scale is lower", maps directly to the central instruction of computers, the "if" statement, as well as to Boolean, On/Off, and reactive control systems. Here again, from the MSC perspective deliberation can be considered as the regular operation of cognitive systems, and is fully described by input – output information flows ("behavior") and time; thus all derived cognitive entities (knowledge, expertise, complexity, etc.) are equally applicable. In this sense, "decision making" or "data processing" other words or expressions that are commonly used, may also more or less loosely be taken as synonyms for deliberation, and thus no new concept and units are required for them.

Top-down approach. A top-down approach implicitly refers to a representation where multiple elements cooperate in a global system, in a hierarchical pyramid. In the MSC model, considering multiple elements (agents, subsystems), there is no such exact notion of higher or lower levels (top-bottom); the MSC model is equally applicable at all granularity scales, i.e., for the overall, integrated system, for the top elements as well for any lower level subsystems. Nevertheless, in MSC, three notions primarily relate to interaction between agents and thus may also partly overlap with the concept of top-down approach: "input–output" information flows, "abstraction–concretization" processes, and "integral systems versus more elementary subsystems".

1. In the first case, the distinction and complementary aspects of input –output flows are obvious, and apply symmetrically for each of the two communicating elements; yet if the information does flow in a single direction, then one element is necessarily a pure transmitter and the other one a pure receiver. Schematically, top-level

elements transmit information, and bottom elements receive it. Examples in the internal control hierarchy of a cooperating robot could be A. the control of stepper motors, or B. the synthesis of speech.

2. Abstraction is the property of cognitive agents that generate less information than they receive; conversely, concretization generates more information (in quantitative terms, abstraction is estimated as the ratio of input information quantity with respect to output information quantity; and concretization as the inverse of the latter ratio). In general, top-down organization calls for concretization. The previous examples also apply here: A. the higher-level, coordination level for motion control receives less information from above (global target values, parameters for motion law) than it generates (interpolated low-level intermediate targets, at higher rates); B. the speech synthesis unit receives less information (text encoding) than it finally generates (CD quality sound waves). Experience shows, however, that the correlation between the top-down approach and concretization is not absolute; high performance cognitive approaches often rely, in some steps, on opposite strategies, e.g., temporarily trading degrees of abstraction for improvements in fluency. For example, expert sorting algorithms include the use of hash tables; or in a subsumption architecture, top-down command components may be cancelled by lower-level reactions.

3. As mentioned above, the MSC model is equally applicable at all granularity scales. This means that if a system is analyzed as a set of subunits, for each subunit the same scheme is applicable; similarly, if several cognitive systems are considered in an integrated, synthetic way, the resulting "meta-system" can also be represented with the same scheme. The notion of hierarchy is therefore "orthogonal", or independent of it (by analogy, notice that in a human group, social hierarchies are in general not apparent in the infrastructure, e.g., human individuals, communication and transportation networks; for another example, the usual objects in a program are similarly defined no matter where they lie in a hierarchy). Another challenge is the mesh aspect of interconnections; instead of a simple, unidirectional level axis (implicit in a top-down structure), what is useful in such contexts is the consideration of multiple dimensions, with parallelism and nested loops. The MSC model supports these views.

Bottom-up approaches. The discussion in the previous paragraph, §Top-down, is applicable here mutatis mutandis. In particular, the bottom-up organization in general calls for abstraction. Examples of such bottom-up, high abstraction cognitive processes for a cooperating robot include A. vision and laser-based localization, and SLAM, and B. Speech recognition.

For each of the concepts that follow, we first A. define the concept in the MSC context, and then B. discuss this concept and its definition.

Creativity. A. Creativity is a particular kind of *knowledge* (measuring unit: lin), that features a *concretization* index higher than 1. B. Knowledge and concretization have already been formally defined in the MSC ontology. The specificity of a creative system is thus simply to ensure that there be actually some concretization, i.e., that more (pertinent, domain-relevant) information is generated by the system than the system itself receives. Creativity is a very common feature of cognitive systems. Here are two examples: building a family house, for an architect; or generating a navigation path from the living room to the fridge, for a domestic service robot.

Ingenuity. A. Ingenuity is a particular kind of *knowledge* (measuring unit: lin), i.e., knowledge in a specific *domain*, which contains reality as the input space and information about novel, improved cognitive system(s) (e.g., better knowledge, better expertise, better intelligence, better abstraction, etc.) as the output space. B. In fundamental terms, ingenuity is knowledge, and as such is already defined in the MSC ontology; the specificity of the domain is of a contingent nature, in the same way as 1. knowing a language contains the instances of 2. knowing French or 3. knowing English, ingenuity is a particular kind of knowledge, involving here a fourth specific cognitive domain. In particular, ingenuity may be considered as the most desirable and prestigious element in the following list of mostly existing MSC concepts related to the generation of information: expression (just generating any sort of information), knowledge (expressing correct information), expertise (doing it correctly and quickly), intelligence (increasing expertise), creativity (knowledge with a concretization index higher than 1, re. previous paragraph), ingenuity (a kind of meta-intelligence by which the cognitive system itself is reengineered so as to yield a quantum improvement in its expertise) .

Chance. A. Chance is a particular kind of *knowledge* (measuring unit: lin), i.e., knowledge in a specific *domain*; the domain contains no input space but an output space consisting of totally unpredictable information. B. In fundamental terms, chance is knowledge, and as such is already defined in the MSC ontology; the specificity of the domain is that it consists of an output space that is totally stochastic. In quantitative terms, in as much as the information delivered is purely stochastic, of potentially unlimited size, the quantity of knowledge here is infinite! As a consequence, in practice, to engineer chance, i.e., to design a perfectly stochastic source, is impossible. Pragmatic solutions, which are satisfactory for some applications, include *approaching* chance with finite resources (e.g., pseudo random generators), and very often *"redirecting"* information generated by an external, natural, source of random information (chance).

Conclusions about new concepts in MSC. In summary it appears that the core elements of MSC model offer a strong basis for cognitive theories, and are quite universal; in the same way as all electronic circuits are made from a very few basic blocks (resistor, diodes, etc.), all logic circuits could be made of NAND gates, or one could argue that virtually all texts could be made out of 26 letters, 128 ASCII characters, or 2 to 4 Morse encoding moments. There are a wealth of other concepts debated in the world of cognition and cognitics. They cannot be ignored and therefore should also be formally defined, as derivatives and sometimes simply special cases of existing, well-defined, core concepts (domain, model, information, knowledge, expertise, etc.). In mathematical terms, we could conclude that the number of concepts commonly discussed far exceeds the actual dimensionality of the space.

4 Generic Schemes for Cognition and Control

We progressively consider five generic schemes for the control of systems, from more established ones to more recent and ambitious ones, where hybrid control as well as automated cognition, i.e., cognitics, play critical roles. Classical elements are listed here along with other, original ones, thus allowing for a more complete perspective.

First, reactive systems are addressed; then, second, improved control, with cognition-based estimates along time and other, partly out-of-reach dimensions are considered; third, the relevance of learning for better results, rather than just relying on expertise, is discussed; then the advantages of temporarily considering subsystems as non-autonomous are mentioned, leaving open the question of whether external components can be integrated at a larger scale; lastly, the use of chance is discussed in cases where ingenuity might seem to be required.

4.1 Reactive Systems

Many control systems are effective and work exclusively on the basis of what is known a priori of the systems to be controlled.

Yet when unknown elements, in particular, disturbances, have a significant impact on systems, it becomes necessary to adopt another scheme. Specifically, it is then useful to incorporate a feedback scheme from the systems being controlled. In subsumption architecture schemes, an additional benefit of going reactive is to avoid the problem of modeling, which a priori knowledge typically implies.

In such circumstances, i.e., considering reactive systems, the primary concern is the relative agility, including input and output communication delays, of the controlling agents, compared to that of the systems to be controlled. Beyond a critical value, control becomes unstable or impossible. (The agility is defined here as the inverse of the global effective reaction time; the critical lower bound on the relative controller agility typically has a value of 2).

T : controller decision time and communication delays
τ : time constant of system to be controlled

Fig. 5. In closed loop systems (e.g., for a simple on/off current limiter in a motor, or for any single dynamic loop of a complex multi-agent, distributed system such as a soccer team), the controller agility (1/T), relatively to the agility of controlled system (1/τ) is critical for success. While above a ratio of 20 control is trivial (an on/off strategy is effective), below a ratio of 2, no solution at all is directly feasible, and reengineering is necessary: either the controller must be made faster, or the system to be controlled made physically slower, or otherwise, one or several additional, ancillary controller(s) is (are) required, thus restructuring the whole as a hierarchy.

4.2 Cognition for the Virtual Exploration of Inaccessible Dimensions and Parameter Regions

In the context of control and automation, cognition may help in several regards. Cognition supports modeling and in particular suppresses the need for many measurements, not only along the time axis (in the future) but also virtually in all dimensions considered.

As introduced above (§4.1) a sufficient amount of application-dependant, a priori knowledge can support open-loop control.

Time. A good knowledge of the controlled system allows for forecasting its behavior, and by this token, a cognitic controller may in principle compensate for some or all control loop delays. This paradigm commonly allows for improvements in effective agility, and consequently commonly brings significant improvements in performance.

Localization. In a subsumption scheme, a robot sensor is located at any given time in a single place. If at the same time, similar estimates of the sensor would be useful at other locations for control purposes, then modeling, and more generally a cognitic approach, can help.

Other dimensions. The power of cognitics is not bound to extrapolation in time and space. On the contrary of situated automata, cognition can extend estimation possibilities in domains out of physical reach. It is in principle quite universal and may often prove similarly useful in almost all other physical dimensions and more. For example, ABB robots accept a weight parameter in their "Grasp" instruction, thereby allowing for updates in the dynamic model of the arm, and ultimately improving their performance.

Fig. 6. Our robot line of service robots in a domestic context, RH-Y (here, RH4-Y in Robocup world competition, in Graz, Austria, 2009) consists of complex systems, featuring classical controllers (PLC, elaborated servo-controllers), computer and processors (tablet PC, etc.) and software solutions (Windows, Linux), as well as novel components, such as our multi-agent, real-time and simulation environment (Piaget), multiple cognitive capabilities such as sensing distances in 3D, recognizing scenes, gestures and speech, locating objects, learning new tasks and topologies, and generating coordinated motions and trajectories as well as producing their own speech and graphics, following humans and handling objects (Re. [3] and numerous videos at http://rahe.populus.ch)

4.3 About Learning

Learning capability seems to be very important for cognitive systems, and indeed learning is at the core of the definition of intelligence in the MSC model. Nevertheless, per se, learning may not be the most useful cognitive property. What is ultimately useful is expertise: the ability to do things correctly and quickly.

Indeed, there are cases where expertise may gradually result from time-consuming learning phases, and similarly expertise must be acquired by each of multiple agent individually, through its own experience. In other cases, the expertise of an agent may also result from some kind of engineering where learning previously performed by another agent has been encapsulated and is directly reused and put to work with minimal delays. Expertise may even suddenly result from novel ideas and ad hoc design, without any learning at all. In all cases, what is finally desirable is expertise.

Consider, for example, human skills: people will typically rely, for advice, on a confirmed expert rather than on a young trainee, even if the latter can learn very fast. The same applies in cognitics: for an elementary task such as multiplying two floating numbers, a circuit at peak expertise level is in principle highly preferable to a possible, in-progress learning system that is gaining experience in the multiplication domain.

It might be argued that sometimes a domain is not stationary enough to allow for the reuse of existing expertise, and that again and again, some learning must be performed there in order to keep a certain level of expertise. However, in such cases there is a very real danger that contradictory requirements may arise: learning is obviously impossible without some stationarity of the related cognitive domain.

4.4 Limited Relevance of External versus Internal Discrimination and Consequence on Estimation of Autonomy

One common question relates to the degree of autonomy of systems. For example in its basic formulation, the task of a neural network learning according to Hebb's law requires that a set of examples, including right answers, be given by a tutor. As another example, consider an elementary controller where target values and possibly feed-forward contributions are given by external agents. As a third and final example, consider a robot following a guide; this requires a guide.

In practice, the development of new concepts and systems usually requires an incremental approach. Similarly, an analytical approach calls for considering, at a given time, only a fraction of the possible overall systems.

In the three cases given above, the current focus leaves one (or possibly many) components as external complements. The system in question must consequently be qualified as non-autonomous. However, as developments progress, or as the analytical scheme moves attention to other components, it may often appear that complements previously considered as external agents are effectively internal to a larger system, and the latter thus may rightfully be recognized as autonomous.

4.5 Chance and Statistical Redundancy as Antidotes to Random Factors

Our guts detect ingenuity in numerous cases (most classical ones include Archimedes's fluid mechanics, Newton's gravity law, or Beethoven's symphonies). Yet the advocated, quantitative approach in the MSC context, say, in "quantitative cognitics", while supporting the incredible merits of expertise as a cognitive property, is still undecided as relates to ingenuity (re. §ingenuity at the end of Section 3).

In quantitative cognitics, a rational explanation for what looks like ingenuity lies in the nature of the following process: the cognitive universality of chance randomly generates novel, revolutionary, finite-size models. Then experts collect and evaluate those models, select the best performing ones and also usually optimize them, in the direction of a very specific goal to approach, and (increasingly, in so far as a related culture exists) with the help of tools and techniques.

The random nature of the source and the exceptional nature of occurrence of adequate models explain the long time usually required for invention; on the other hand, some redundancy in attempts to search for a model of that kind, and good expertise levels of resources engaged in collecting, selecting and optimizing novel models, improve the probability of success.

5 Conclusion

Intelligent systems and robots are continually progressing. Several components have been especially responsible for this development: AI, situated automata approach, closed-loop control, computer infrastructure and networks. Nevertheless limitations remain, which can be pushed farther, though not eliminated, by hybridization and a quantitative cognitics approach.

Cognitics, i.e., automated cognition, and hybrid approaches offer multiple benefits. In cognitics, the MSC model has been developed, which includes an ontology and a metric system; this provides a sound basis for assessing and complementing the various elements contributing to current solutions. Hybrid approaches integrate the specific advantages offered by classical AI (predict phenomena), control (compensate for disturbances), reactive systems (embed in reality so as to simplify representation), cognitics (the move toward quantitative and automated systems in the cognitive world) and computer technologies (implement on digital processors, with the best of software methods). In addition, ingenuity, chance and time are brought into the general picture.

Hybrid approaches, and specifically the RAS-HYCAS attempt, inevitably bring together analog notions that may collide, in the sense that, stemming from different contexts, their meanings do not always perfectly match. Notions like deliberation, control, reactive systems or top-down approaches are therefore newly discussed from the MSC perspective, while other concepts, in particular, creativity, ingenuity and chance, are added to the MSC ontology.

Generic schemes have been presented above in which classical control as well as reactive systems are very usefully complemented by AI, and more generally cognitive approaches: closed-loop control is critically sensitive to time aspects (relative agility of controller), reactive systems essentially restricted to "here and now", while cognition can compensate for delays and estimate entities in virtually all regions of high dimensionality spaces; intuitively, learning (and intelligence) often appear to be important while in fact *expertise* has typically proven to be the decisive cognitive property in the MSC context; the difference between external versus internal discrimination is often more a matter of time, or a single step in an ongoing process of development, than a fundamental difference; finally, an expert process harnessing chance is a more rational source for radical innovation than a hard-to-define, intuitive notion of ingenuity.

In summary, in the quest for ingenuity, a quantitative approach demonstrates that no matter how complex a model is, it is always infinitesimal in terms of informational content compared to the amount of information necessary to describe the underlying reality; consequently, only chance has the cognitive power to yield novel models. Nevertheless, it can also be argued that precisely because a novel, vastly improved model is a rare phenomenon (there is an essential relation between high information and low probability of occurrence), it is of utmost importance that, whenever this happens, expert resources collect it, optimize it, and make use of it for the benefit of mankind.

References

1. Brooks, R.A.: Intelligence Without Representation. Artificial Intelligence 47, 139–159 (1991)
2. Dessimoz, J.-D., Gauthey, P.-F., Kjeldsen, C.: Ontology for Cognitics, Closed-Loop Agility Constraint, and Case Study in Embedded Autonomous Systems – a Mobile Robot with Industrial-Grade Components. In: Proc. Conf. INDIN 2006 on Industrial Informatics, August 14-17, p. 6. IEEE, Singapore (2006)
3. Dessimoz, J.-D., Gauthey, P.-F.: RH4-Y – Toward A Cooperating Robot for Home Applications. In: Robocup-at-Home League, Proceedings Robocup 2009 Symposium and World Competition, Graz, Austria, June-July (2009)
4. Ferrein, A., Pauli, J., Siebel, N.T., Steinbauer, G.: Preface and Proceedings, 1st International Workshop on Hybrid Control of Autonomous Systems — Integrating Learning, Deliberation and Reactive Control (HYCAS 2009), Pasadena, California, USA (July 2009)
5. Pi-search webpage (November 10, 2009), http://www.angio.net/pi/piquery
6. Shannon, C.E.: A mathematical theory of communication. Bell System Technical Journal 27, 379–423, 623-656 (1948)
7. Turing, A.M.: Computing Machinery and Intelligence. Mind, New Series 59(236), 433–460 (1950)

Appendix A

In support of the brief considerations summarized in §2.3 above, and as another example of the potential benefits of hybrid approaches, let's revisit the striking case of the accurate estimation of the number "Pi," 3.14... Computer technology is here a necessity; and yet reciprocally, a quantitative approach quickly demonstrates that cognitive processes can be a very powerful complement to mere computer technology! Pi can simply be defined as an infinite series as follows: $Pi=4*(1-1/3+1/5-1/7...)$.

A standard, double precision number system yields, say, 40 significant digits. Let us very optimistically assume we can compute numbers with arbitrarily large numbers of digits, and that an elementary operation (denominator update, inverse and subtraction) takes 1 nanosecond on a computer. The accuracy of the estimate is here on the order of the number of operations, the nth element being worth about $1/n$. By this token, after 1 second of operation, $n=10^9$, so we have an accuracy of about 10^{-9}, i.e., this yields about 9 significant digits. The problem is that improvements are very slow, logarithmically slow: after one year of computation, i.e., 10^7 additional seconds, we would have 7 more significant digits, i.e 17 of them in total; and if the computer could already have been computing since the Big Bang, i.e., 10^{10} years ago, we would now have a mere 10 additional significant digits, thus a total of 26 digits! The same applies for technology: assume that computers could get a billion times faster, we would only get 10 more digits from this series. Meanwhile, cognition, or we can loosely say, ingenuity, is unbelievably powerful: twenty years ago, 4000 digits had been discovered; by 2001, 100 million digits, and by 2005, 200 million digits [5] were accurately identified!

Mechanical Design and System Architecture of a Tracked Vehicle Robot for Urban Search and Rescue Operations

Raimund Edlinger, Andreas Pölzleithner, and Michael Zauner

Upper Austria University of Applied Sciences,
Research & Development,
Stelzhamerstrasse 23, 4600 Wels, Austria
{raimund.edlinger,michael.zauner}@fh-wels.at
http://www.fh-ooe.at/en/rd/forschung/

Abstract. Human rescuers have very short time to find trapped victims in a collapsed structure, otherwise the chance of finding victims still alive is nearly zero. In such a critical situation robots can support rescuers. According to this very demanding tasks the development of the robots is very complex and combines multiple disciplines such as mechanical engineering, electrical engineering and programming. This mobile platform equipped with a manipulator represents the first stage of an Austrian national founded project. A common background of our above activities is the development of a tracked vehicle with four active flippers to improve the traversability on uneven terrains, the control of the driving mechanism and the system architecture with the experiments in the programming of an embedded real-time processor and a high-performance FPGA from *National Instruments* to control all sensors and actuators.

Keywords: Rescue Robot, tracked mobile robot, stair climbing, mobile manipulation.

1 Introduction

Intelligent mobile robots and cooperative multi-agent robotic systems can be very efficient tools to speed up search and rescue operations. Rescue robots are also useful to execute rescuing jobs in situations that are hazardous for human rescuers. They can enter gaps and move trough small holes that are not accessible for humans or even trained dogs. Robots are supposed to explore collapsed buildings, extract the map, search and locate victims in map and way that rescue teams can reach the victims.

It requires the highest demands on the motor and sensory abilities of the robots. Figure 1 shows the result of previous works on tracked mobile vehicle in particular for search and rescue operations. The robot is developed specially for application in the field of security and emergency and is supposed to act either autonomously or remote controlled by humans.

D. Obdržálek and A. Gottscheber (Eds.): EUROBOT 2010, CCIS 156, pp. 46–56, 2011.

1.1   Specification

For the development of the robot the following specifications were defined:

- The robot is either controlled autonomously or remote control via wireless communication;
- Finding victims by using sensors such as a vision system, CO_2 sensors, microphone, etc.
- Localizing source of danger
- Driving in uneven terrain for example stairs, unstructured environment
- Generating practical maps of the environment

2   Related Works

There is a number of researchers that have created different mechanism, techniques, and methods to explore uneven terrain by using mobile robots. Several mobile platforms have been developed with four active flippers [1] and [6], robots with a simple additional flipper with one active DOF [1] or the development of wheeled mobile robots [5] which participate in RoboCup Rescue League [2]. An other mechanism for the mobility is realized in [4], which is equipped with a tracked base and a manipulator for handling operations and for support of the vehicle mobility. Industry companies [7], [8] and [9] produce robots that have such mechanism with an effectiveness to adapt unstructured environment. The introduced robot is the first robot prototype of upper austria university of applied sciences, see fig. 1, of the RoboRescue Team [13]. It is a tracked vehicle equipped with flippers which are very important to improve traversability on uneven terrain. During the remote controlled modus the motion of the robot is controlled by a *Logitech* force-feedback joystick which is connected to the operator station.

Fig. 1. The old and new version

3   Mechanical Design

This section explains the main features of the mechanical design of the robot and the dimensions of the robot which is shown in Table 1. All the custom parts are machined at the LKR [10] which is a research center with the focus on lightweight constructions.

Table 1. Dimensions of the robot

Track (lenght)	535mm
Track (width)	510mm
Track (height)	510mm
Length (extended)	980mm
Width (extended)	510mm
Height (extended)	1310mm
Weight	42kg
Speed on flat ground	4km/h

3.1   Robot Locomotion

The locomotion of mobile robots in undeveloped outskirt area is one of the most difficult demands on the system. On one hand, as an outdoor robot it has to be fast and flexible on the other hand the vehicle has to deal with rough underground such as stones, gravel or stairs. Other important requirements are that the whole system is, on one hand robust, and on the other hand a lightweight

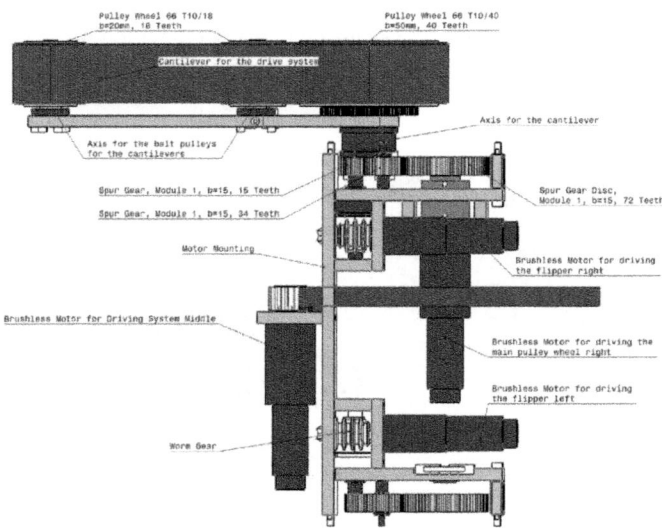

Fig. 2. Mechanism of the driving system

construction to reduce the energy consumption and increase the agility. According to these requirements a drive system is developed which is shown in fig. 1. One active flipper consists of two brushless motors; one motor drives the main pulley wheel, the second one is supporting the cantilever which is shown in figure 2. The drive system basically consists of four pulley belts which are driven separately. Additionally the two belts on the left and right side and the middle belts can rotate individually. This is important for tasks like driving over uneven underground and climbing stairs. The body of the vehicle basically consists of an aluminum frame and the gaps, which are for reducing weight, are covered with carbon composite sheets.

The brushless motors and the gear boxes are located inside of the frame so they are protected against damage and pollution.

3.2 Manipulator [14]

A very important feature is the robotic arm which is mounted on the top of the vehicle. One thermo camera two standard cameras and several other sensors are attached as payload at the very end of the arm. This makes it possible to reach hard accessible areas up to approximately 120cm above ground including the vehicle beneath. It is designed to carry a payload of approximately 2kg. The robotic arm has six degrees of freedom which are driven by DC-motors via worm gears. For reducing weight and increase the stability modern materials such as special aluminum alloys, magnesium and carbon composite materials are used. The links and joints are specially designed to be able to change the configuration of the arm easily by adding or removing links if it is required. Furthermore the bodies of the main links are realized by carbon composite tubes which can be replaced by tubes with different length. This makes it a modular system. The measurement of the angular position of each joint and a forward kinematic model allows determining the position and orientation of the camera at the very end. For the next iteration an inverse kinematic model and a trajectory planning algorithm are planned.

Functional Design. The robotic arm consists of an baseplate, five rigid links, the end effector at the very end and a flange to mount the payload. The links are connected to a kinematic chain via six joints which are all rotary. This gives six degrees of freedom. Figure 3 shows a schematic drawing of the robotic arm. The first degree of freedom allows the whole robotic arm to turn around an vertical axis. The following three degrees of freedoms are around vertical axis and connect the links that are actually lifting the payload. The two last axis are carried out like a human wrist.

The whole robotic arm is designed using the CAD[1] program CATIA from *Dassault Systems* which allows to draw a three dimensional model. The drawings can be exported for simulation and machining. The arm is assembled out of custom made aluminum and magnesium parts, standard parts such as ball bearings and carbon fiber tubes.

---

[1] Computer Aided Design.

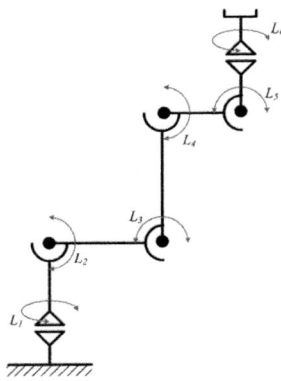

Fig. 3. Schematic of the kinematic chain

Links and Joints. As mentioned the arm is connected to the robot by a baseplate. The second link is U-shaped and is between the first and the second axis. The main links consist of carbon fiber tubes with pivot elements on both sides which are made out the aluminum alloy EN AW 7075. An pivot axis is mounted on the fixed part of the pivot which is rotative supported by ball bearings in the opposite part of the pivot. The driving motors are located inside of the tubes and drive the joints via a worm gear. Aluminum inserts are glued into both ends of the tubes to mount the pivot parts. As mentioned the hinges are designed to achieve a modular system. This is realized by using the same hinge type for each joint. Figure 4 shows an explosion drawing of the joint.

This type of hinge is applied for every joint except for the first and the last. The very last link is very complex because function is similar to the function human wrist. The last axis is orthogonal to the second last which allows the end effector to yaw and pitch. In order to reduce weight, the fifth link is made out of magnesium. An explosion drawing of the last two links is shown in figure 5.

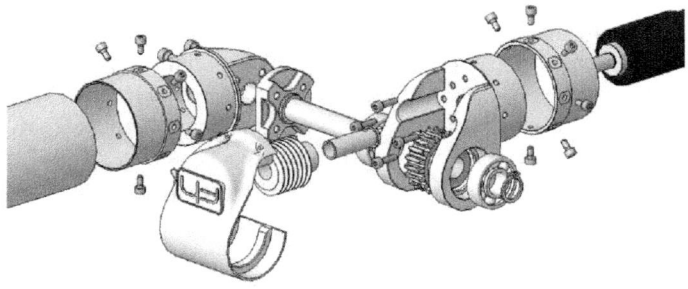

Fig. 4. Explosion drawing of the main joint

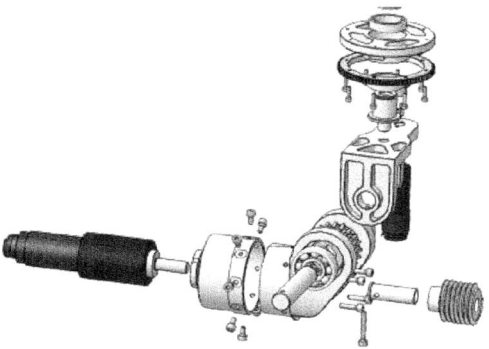

Fig. 5. Explosion drawing of the last two links

Driving systems. For driving the joints DC-motors with a planetary gear head and magnetic encoder are used. Except the last all joints are driven via worm gears so there is no power consumption in the static state. The last joint is driven via bevel gears. Encoders which are mounted on the rear side of the motors are used to measure the angular position of each joint. These encoders are also used to get feedback for the discrete negative feedback controller of the motor driver units. Since the encoder does not provide absolute values, opto-coupler is used to acquire a reference angle. The encoders have a resolution of 16 steps per revolution. Due to a transition ratio of 1 : 575 the angular accuracy is 0:039 per step.

Finite element simulation. To ensure that all parts are designed properly the robotic arm is simulated with the finite element simulation program ANSYS, see fig. 6. For the simulation the CAD drawings are used. The simulation gives basic information of stress and pressure in the materials. Furthermore the deformation by its own weight and an additional payload is computed [10]. The made it possible to assign the appropriate material to each part.

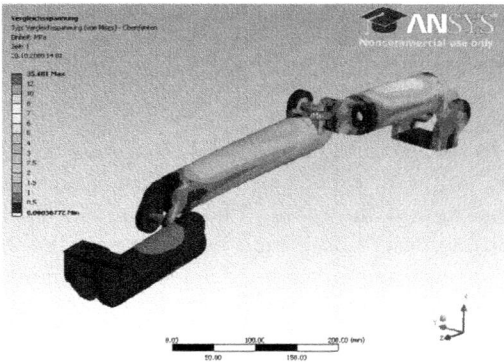

Fig. 6. Simulation of the robotic arm using ANSYS

Kinematic Model - Denavit-Hartenberg Notation. A kinematic model is calculated to determine the position and orientation of the end effector out of the axis variables which are the angular position of each joint. To set up the forward kinematic model vector spaces and transformation equations are used. Therefore a homogeneous coordinate system is assigned to each link which are all considered as rigid. These assigned coordinate systems are named frames and define the orientation and position of each link in respect to the former.

A convention, developed by J. Denavit and R. S. Hartenberg [15], defines a method to assign the frames to the links in a useful way. This method is introduced in [11]. The assigned frames are the transformation matrices end are generally denoted as in (1).

$$
{}^{i}_{i+1}T = \begin{pmatrix} cos\theta_i & -\lambda_i sin\theta_i & \mu_i sin\theta_i & a_i cos\theta_i \\ sin\theta_i & \lambda_i cos\theta_i & -\mu_i cos\theta_i & a_i sin\theta_i \\ 0 & \mu_i & \lambda_i\theta_i & b_i \\ 0 & 0 & 0 & 1 \end{pmatrix}
\tag{1}
$$

The variables a_i, b_i, μ_i and λ_i are the "Denavit-Hartenberg parameters" which depend on the mechanical construction of the respective link and the axis orientation. The variable θ_i is the axis variable and represents the actual angular position of the respective joint. In this case six transformation matrices are assigned. The position and orientation of the end effector in respect to the first frame can be computed by multiplying the transformation matrices.

$$
{}^{1}_{e}T = {}^{1}_{2}T \cdot {}^{2}_{3}T \cdot {}^{3}_{4}T \cdot {}^{4}_{5}T \cdot {}^{5}_{6}T \cdot {}^{6}_{e}T
\tag{2}
$$

In this way the orientation and the position of the end effector can be calculated by the controller.

4   System Architecture

4.1   Controlling System

As main control system of the robot the SINGLE BOARD RIO from *National Instruments* is used. It is a modular controlling system which in charge of monitoring the sensor data, controlling the actuators execution the main task for planning and navigation of the robot. The NI SINGLE-BOARD RIO integrates an embedded real-time processor, a high-performance FPGA, and onboard analog and digital I/O on a single board. All I/Os are connected directly to the FPGA, providing low-level customization of timing and I/O signal processing. The FPGA is connected to the embedded real-time processor via a high-speed PCI bus [16]. The basic architectures for real-time mission control are summarized in figure 7.

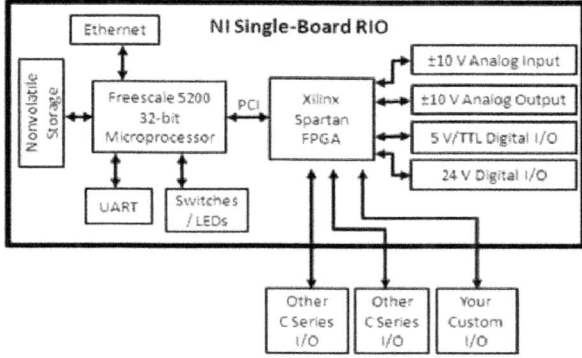

Fig. 7. NI Single-Board RIO Architecture [16]

The full range of employment requires a distributed system architecture by using different bus systems. The advantage of this architecture is the modular and flexible implementation of tasks and the use of global function variables which are programmed in the main application framework:

- The *Main Routine*, which is implemented in the *NI-SBRIO*, controls all activities of the robot and receive/send information to the operator.

- The *Motion Task*, which is also realized in the *SBRIO*, manages the motion of all 14 motors to encourage a stability of the vehicle and to ensure an obstacle avoidance of the manipulator.

- The *Communication Task* stand for the wireless LAN communication to the host computer.

- The *Vision Task* represents the vision system with stereo metric.

- The *Control Task* ensures the remote control of the rescue robot by a lightweight laptop (HP Compaq 8510p) via a LOGITECH force feedback joystick.

- The *Mapping Task* guarantees a two dimensional map generation based on the acquired data of a laser range finder *LRF UBG-04LX-F01* from the company HOKUYO [17].

4.2 Motion Control

The *Solo-WHI*, which is developed by ELMO MOTION CONTROL, was designed to be a very compact product oriented for X, Y, Z applications integrating three servo drives in one package and working under one power supply equipped by a built-in shunt regulator. The Trio operates from an AC/DC power source in current, velocity, position and advanced position modes, with permanent-magnet synchronous brushless motors, DC brush motors, linear motors or voice coils. A CANOPEN network connects the servo drives to the main controller which is the NI SINGLE BOARD RIO.

5   Victim Detection

For victim detection a thermo camera and one standard camera are mounted on the top of a robotic arm and provide pictures for the operator. A graphical user interface (GUI) is about to be developed which is supposed to display current information of the terrain and environment. Furthermore the GUI supposes the sensor data of the CO_2 sensor, laser range finder and several other sensors. Additionally the operator gets important information about the robot's battery status and warnings for the obstacle avoidance. The two dimensional map generation is based on the acquired data of a laser range finder (LRF) UBG-04LX-F01 from the company *Hokuyo* [17]. The light source of the sensor is an infrared laser with a wavelength of 785nm (laser class 1). The sensor is connected to the USB2.0 port of the MiniITX and communicates with the map building software which is programmed in C++. For generating the map the *scan-matching* algorithm ICP-Iterative Closest Point and *Coreslam* [12] combined with landmarks is used, see figure 8. The laser range finder obtains data of a scanning angle of 240 and a maximum distance of 4m. It is mounted on the front side of the robot and is automatically leveled out by a mechanism in case that the robot tilts.

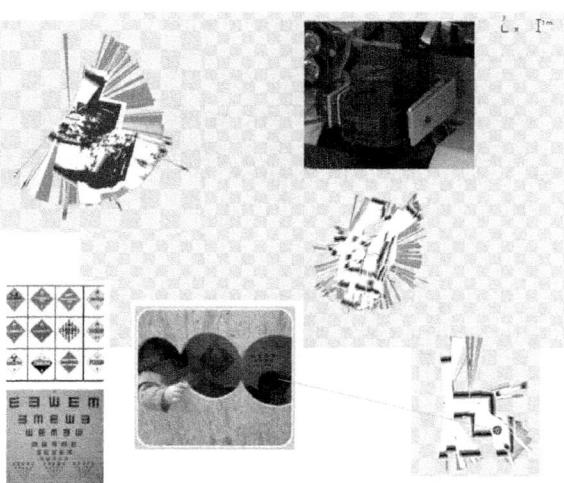

Fig. 8. The autonomously generated map with ICP-Scan Matching

In order to detect position and attitude of the robot an inertial measurement unit, so called IMU, is used. The IMU consists of three accelerometers, to determine the acceleration in the three axes and three rate sensors, to detect the attitude. Additionally three magnet sensors are used to achieve the direction of the earth magnetic field for referencing the sensor. For illumination two LUX-EON LED rings are used. It also allows analysing the light of the environment around the robot. In order to get a constant illumination level the brightness of the LEDs is controlled. This gives the advantage of a very stable and constant histogram which is important for an automatic detection of insured people.

6   Conclusions

6.1   Conclusions

Rescue robots are making the transition from an interesting idea to an integral part of emergency response. Rescue robots present challenges in all major subsystems (mobility, communications, control, sensors and power). The new rescue robot from the *Upper Austria University of Applied Sciences* has been introduced. The mechanical design, concept configuration and system architecture was explained. One of the big features and novelty is the use of the *SBRIO* from *National Instruments* and the motion control from *ELMO*. This robot is a polymorphic tracked vehicle represent the state of the practice of ground rescue robots. The tracked vehicle has been primarily used for navigation and sensing missions, that require manipulation. The advantage of the robot is, that it extend the capabilities of ground vehicles by allowing the robot to sample the environment, interact the survivors, add unique camera views and move light obscurations. The experimental results show that the remote teleoperative robot has the mobility to overcome obstacles, continuous up to 15 ramp terrain, 40-45 stairs and step field pallet terrain.

6.2   Future Works

The aim is to improve the performance of the strategy of autonomous motion of flippers for the tracked vehicle and the autonomous robot navigation and victim identification. In the future the robot is supposed to be extended by a grabber for pick and place task such as manipulating radios, sensor package or a water bottle. It is also planned to program a globally consistent representation (3D-Mapping) of a robot's environment for better localization and navigation.

Acknowledgment. The authors gratefully acknowledge the contribution of the *Upper Austria University of Applied Sciences* who is sponsoring this research and the contribution of Ziad Khalil and the team of the "ARC Leichtmetallkompetenzzentrum Ranshofen" for sharing their knowledge, longstanding experience and support in realizing the project.

References

1. Birk, A., Pathak, K., Schwertfeger, S., Chonnaparamutt, W.: The IUB Rugbot: an intelligent, rugged mobile robot for search and rescue operations. In: International Workshop on Safety, Security and Rescue Robotics (SSRR). IEEE press, Los Alamitos (2006)
2. Jacoff, A., Messina, E., Weiss, B.A., Tadokoro, S., Nakagawa, Y.: Test arenas and performance metrics for urban search and rescue robots. In: Proc. of the IEEE/RSJ International Conference on Intelligent Robots and Systems, pp. 3396–3403 (2003)
3. Nagatani, K., Yamasaki, A., Yoshida, K., Yoshida, T., Koyanagi, E.: Semi-autonomous Traversal on Uneven Terrain for a Tracked Vehicle using Autonomous Control of Active Flippers. In: Proc. of the IEEE/RSJ International Conference on Intelligent Robots and Systems, pp. 2667–2672 (2008)
4. Guarnieri, M., Takao, I., Debenest, P., Takita, K., Fukushima, E., Hirose, S.: HELIOS IX Tracked Vehicle for Urban Search and Rescue Operations: Mechanical Design and First Tests. In: Proc. of the IEEE/RSJ International Conference on Intelligent Robots and Systems, pp. 1612–1617 (2008)
5. Pellenz, J., Koblenz, U.: RoboCup Rescue League Teampaper. RoboCupRescue Team resko@UniKoblenz, Germany (2008)
6. Dornhege, C., Kümmerle, R., Ruhnke, M., Steder, B., Nebel, B., Doherty, P., Wzorek, M., Rudol, P., Conte, G., Durante, S., Lundstrom, D., Kleiner, A.: RoboCup Rescue League Teampaper. In: RoboCupRescue - Robot League Team RescueRobots University Freiburg, Germany (2006)
7. I-Robot Ltd., "Packbot", (October 2008), http://www.irobot.com/
8. Telerob Ltd., "Telemax" (October 2008), http://www.telerob.de/-OnlineResource
9. Robowatch Ltd., "Asendro" (October 2008), http://www.asendro.de/index.php?id=110&L=0-OnlineResource
10. ARC Leichtmetallkompetenzzentrum Ranshofen Ltd., "LKR" (February 2008), http://www.lkr.at/-OnlineResource
11. Angeles, Jorge.: Fundamentals of Robotic Mechanical Systems: Theory, New York, Inc., pp. 265–271 (1995)
12. CoreSlam, Simultaneous Localization And Mapping (June 2009), http://www.openslam.org/
13. RoboRescue Team, Research & Development Department, University of Applied Sciences Upper Austria (June 2009), http://rrt.fh-wels.at
14. Poelzleithner, A.: Diploma Thesis: Robotic Arm for an Autonomous Rescue Robot, University of Applied Sciences Upper Austria (June 2009)
15. Denavit, J., Hartenberg, R.S.: A kinematic notation for lower pair mechanisms based on matrices. ASME (1955)
16. National Instruments Ltd., NI-Single Board RIO (March 2010), http://www.ni.com/singleboard/d/
17. Hokuyo Ltd., "UBG-04LX-F01" (March 2010), http://www.hokuyo-aut.jp/02sensor/07scanner/ubg_04lx_f01.html

Flexible Robot Strategy Design Using Belief-Desire-Intention Model

Loris Fichera[1], Daniele Marletta[2], Vincenzo Nicosia[3], and Corrado Santoro[4]

[1] University of Catania – Engineering Faculty
Viale Andrea Doria, 6 — 95125 - Catania, Italy
loris.fichera@gmail.com
[2] Scuola Superiore di Catania
Via S. Nullo 5/i — 95123 – Catania, Italy
danielemar86@gmail.com
[3] Scuola Superiore di Catania – Laboratorio sui Sistemi Complessi
Via S. Nullo 5/i — 95123 – Catania, Italy
vincenzo.nicosia@ct.infn.it
[4] University of Catania – Dept. of Mathematics and Computer Science
Viale Andrea Doria, 6 — 95125 - Catania, Italy
santoro@dmi.unict.it

Abstract. This paper describes PROFETA, a Python framework developed by the authors to write robot strategies by means of the Belief-Desire-Intention (BDI) programming paradigm. This paradigm has been proposed in the field of autonomous agents programming and can be successfully applied also to autonomous robots thanks to their behavioural similarity with software agents. The paper describes the BDI model and AgentSpeak, a formal declarative language suitably designed for BDI agents. Then it introduces PROFETA, which takes inspiration from AgentSpeak and is designed with the objective of adding declarative constructs (needed by a BDI model) to an object-oriented and imperative language like Python. The result is a flexible environment that combines the power of both the classical object-oriented paradigm—useful for algorithm and control loop programming—and declarative approach—useful for AI and strategy programming. A case-study, based on Eurobot 2010 competition, shows such abilities, highlighting the main characteristics and advantages of PROFETA in strategy design.

Keywords: belief-desire-intention model, autonomous robots, artificial intelligence, python.

1 Introduction

The design and implementation of the software running on autonomous mobile robots (AMR), like the ones participating to the Eurobot competitions, requires to face and solve a lot of issues related to different areas of computer science. While the control loops for mechanical arms or speed and motion control are often implemented by means of simple C or assembly programs running on microcontrollers, higher level tasks, such as path planning, obstacle avoidance,

D. Obdržálek and A. Gottscheber (Eds.): EUROBOT 2010, CCIS 156, pp. 57–71, 2011.

artificial vision or strategy control, require more sophisticated algorithms and are usually solved by artificial intelligence or soft computing techniques. In order to separate the various tasks which are responsible of controlling the overall behaviour of the robot, and to facilitate the interaction among them, a proper software infrastructure must be designed. In particular, the separation between low–level tasks (actuation and sensing) and higher level and "intelligent" tasks (strategy, planning, reasoning) is often implemented by means of *layered software structures* [1,2,3,4].

The most notable advantage of a layered structure is that a clear separation of the various tasks allows an easy refactoring of the system: adding a new functionality or modifying an existing one does only affect the layer(s) to which the functionality belongs, leaving untouched the rest of the system. Undoubtedly, this gives a great flexibility in the incremental design of the software system of an AMR, and is very useful during the planning and the from–scratch implementation of task–specific autonomous robots, such as those participating to the Eurobot contest, and in general in all the cases when the technical solutions implemented on the robot can sensibly change after the planning phase. Unfortunately, such layered architectures could be not enough in situations where a great flexibility is required in the upper layers, i.e. for the "robot's intelligence". Indeed, a robot playing a game, like those of Eurobot, has to implement a strategy which usually consists in repetitively *(i)* approach/recognize an object, *(ii)* pick the object and store it, *(iii)* put the object in an appropriate place. According to the year's rules, objects can be of different types and thus need to be sorted, put in different containers or in specific sequences to gain more points, etc. Therefore it is important to find a *winning* strategy, which possibly takes into account all the different unexpected situations that could happen during a match.

First of all, even if playing objects are placed in well-known points of the playing area, they may move due to accidental collisions with the robots or with other playing elements, and thus they can be found almost everywhere in the playing field, also in places in which the robot never expects to find them. Similarly, the opponent robot could be found in any place and at any time during the match, and collisions with it must be avoided. Moreover, the strategies for managing unexpected situations or to avoid collisions with the opponent can in principle be different according to what the robot is currently doing (e.g. going to pick an element or trying to deposit it etc.). Another aspect which heavily affects strategy is the limited duration of the game, which lasts only 90 seconds; different actions could be indeed required when the game time is elapsing and it could be better to e.g. try to put the few gathered objects as soon as possible instead of trying to search and pick other objects. Finally, it is desirable to have an easy and flexible way to change the robot strategy for each match, for instance to effectively react to the strategy employed by the opponent.

Developing a proper strategy which takes into account all the aspects mentioned above is not an easy task. From the point of view of software, either an imperative/object–oriented or a logic/declarative paradigm could be used, but sometimes a mixture of both gives better results. The classical imperative approach exploits the finite-state machine (FSM) abstraction, in which an event occurring in a certain state (such as a detected object, the presence of an obstacle on the way, the reaching of a certain point, the elapsing of a timer, etc.) triggers the execution of a specific action and the change of the state of the FSM. Practically, implementing such a FSM results in a source code consisting of statements like *"if (event and condition) then do_something()"*, which generally tends to be quite hard to read, debug and maintain when the complexity of the overall strategy increases.

Another approach exploits *inference-based systems*, which are typical in artificial intelligence applications [5]; the robot is supposed to have a certain *knowledge* about itself and the environment, i.e. it is in a certain position, has gathered some objects, the opponent is close or far away, etc.; such a knowledge is fed to a proper set of *inference rules* which, on the basis of a *knowledge-based deductive process*, allow to infer which are the actions to be performed in the current situation and to acquire new knowledge. Such systems can be implemented by means of languages and tools based on logic/declarative approaches, such as Prolog or LISP (as for the languages) or OPS-5, CLIPS, JESS, Drools [6,7,8,9] (as for the tools).

Indeed, since an AMR usually uses a lot of code written in an imperative (and often low-level) language such as C or assembly, employing a logic/declarative paradigm to implement strategies implies mixing the two programming approaches, which often makes the overall software system very hard to read, debug and change. As pointed out in [10,11,12,13], having different software subsystems implemented with different programming languages, inspired by different paradigms, requires to face many problems, not last the fact that serious efficiency issues arise due to the need to make the imperative and declarative parts interact with one another (i.e. to pass data or commands between high–level logic programs which implement the strategy and low–level firmware which control actuators and sensors).

Starting from the assumptions discussed above, in 2003 the authors' research group started an investigation aiming at studying, proposing and implementing languages and tools able to support the development of efficient and agile software architectures for autonomous systems programming. The main target of this research is to find valuable solutions which avoid the undesirable "mixture of programming paradigms" [10,11,12,13,14,4], which, in the context of AMRs, means to support the implementation of both control loops and strategy logic using *the same* approach. Following this line of investigation, this paper proposes a framework, for the Python language, which help developers in implementing strategies for an AMR, by means of a logic/declarative approach. The framework, called PROFETA *(Python RObotic Framework for dEsigning sTrAtegies)*, is based on AgentSpeak(L) [15], a kernel language to implement software agent systems with the Belief-Desire-Intention (BDI) model [16,17]. By exploiting some peculiar features of Python (mainly object-orientation and flexible operator overloading), PROFETA is able to add declarative constructs to Python programs, while a proper set of classes provides the suitable abstractions to support the specific concepts introduced by the BDI model. As a result, the overall software system of an AMR can be supported by means of a *single environment*, that is the Python virtual machine, where the classical OO/imperative constructs can be exploited for the control loops, while the "intelligence" can be implemented also in Python, but using the additional constructs provided by PROFETA.

The paper is structured as follows. Section 2 provides an overview of the BDI model and the AgentSpeak(L) language, also showing some examples of its usage in autonomous robot programming. Section 3 presents PROFETA, highlighting the basic classes, the syntactic constructs and the functionalities. Section 4 presents a case-study of PROFETA by discussing the peculiar aspects of the strategy employed by the robot built by the UNICT-TEAM and participating to 2010 edition of Eurobot. Section 5 deals with related work. Section 6 concludes the paper.

2   The BDI Model and AgentSpeak(L)

An increasing number of real-world applications run in a dynamic environment and require complex software architectures characterized by an high level of *interaction*. The scenario of the Eurobot competitions is a typical example where the software system of an AMR has to adapt its behaviour to properly manage unexpected situations which can occur during the game. This year's game rules state that the robots have to collect different objects representing corns, tomatoes and oranges, and then drop them in a basket. However, these items may unpredictably move, e.g. because a robot has (un)successfully collected them. Moreover, among the playing elements there are some fake black corns placed in random positions which should not be collected. The robots have also to avoid obstacles which could be found along their paths. Finally, the robots must autonomously employ a good strategy to score as much points as possible and win the game.

Designing strategies for a system that acts in such a complex and dynamically changing environment is an hard task: conventional programming paradigms, based on imperative approaches, do not adequately address all issues that arise when building such highly reactive systems which have continuous interaction with the environment. As a consequence, a lot of effort has been spent in proposing new methodologies and tools explicitly thought to facilitate the development of efficient autonomous systems; the *agent-oriented programming* (AOP) paradigm is currently considered one of the most interesting solutions [18,19].

AOP is based on the concept of *agents* [20,21], i.e. entities that *act* in a dynamic environment, *perceive* their surroundings through sensors, *adapt* their behaviour to respond to changes in the environment and make *autonomous* decisions to achieve their design goals. Given this definition, it is clear that the notion of agent is well suited to model the behaviour of an AMR, which has all the above mentioned characteristics.

Several approaches were proposed to model the agent's internal *mental state*, i.e. the set of components that represent the environment, the state of the agent itself and the logic to select the appropriate actions to execute. In the following, we will describe one of the most successful approach, namely the BDI model. Then we will present an overview of AgentSpeak(L), an abstract language specifically thought to program BDI agents.

2.1   The BDI Model

The BDI model has emerged as one of the most interesting proposals in the agent-oriented literature [16]; based on a philosophical theory of human practical reasoning [17], it assumes that the *mental* state of an agent consists of three fundamental *attitudes*: Beliefs, Desires and Intentions.

Beliefs represent the *informational* state of the agent, i.e. its knowledge about both the external world and the internal state of the agent. Since the environment is dynamic, the beliefs must be properly updated by using information provided by sensors. In our example, the coordinates of the corns, the position of the basket, the current number of oranges or tomatoes picked by the robot, the dimensions of the playground are all beliefs of the robot. Whenever the sensors perceive that a corn is not in the expected location, the beliefs of the robot are updated to register this change in the state of the playground.

Desires, or more commonly Goals, represent the *motivational* state of the agent. In general, they refer to the set of objectives the agent wants to achieve. Picking a corn, avoiding an obstacle, searching a tomato and reaching the basket are all examples of Desires for the robot of our example.

Since the agent acts in its environment, it executes a proper sequence of actions in order to achieve its objectives. The BDI model points out the necessity to explicitly represent this sequence of actions, which is called **Intention** and corresponds to the *deliberative* state of the agent.

The reasoning process of a BDI agent consists in the selection of an appropriate *plan*, that is a sequence of actions or (sub)goals that represents the mean by which an agent is able to achieve a specific objective and/or respond to changes in the environment. In our example, the robot could have a plan to pick four corns in a row, a plan to reach the basket while picking some tomatoes, a plan that gets triggered when it discovers that the corn it was trying to pick is actually black, and so on. The set of all plans represents the *plan library* of the agent. We can note that in general completing a plan is equivalent to reaching a series of intermediate goals and performing a given sequence of actions.

In summary, we can define a BDI agent by specifying its knowledge base and its plan library. The reasoning process of a BDI system is based on the definition, composition, revision and execution of plans. Plans represent a simple and flexible way to model the behaviour of an agent, and by means of this model the mental state of an agent becomes similar to that of the human mind.

Several works were inspired by the BDI model [22,23,24,15,25]. In the following we give a brief introduction to one of the most used BDI languages, AgentSpeak.

2.2   AgentSpeak(L)

AgentSpeak(L), or simply AgentSpeak [15], is an abstract framework that uses the logic/declarative approach and the BDI model to define rational BDI agents.

In AgentSpeak, a belief is simply written as a ground atomic formula. For example, `corn(150,722)` and `tomato(600,1222)` are belief atoms that represent the coordinates of a corn and a tomato respectively in the robot's knowledge base.

AgentSpeak defines two categories of goals: *achievement* goal and *test* goal. An achievement goal is denoted by the operator '!' followed by an atomic formula. It specifies that the agent wants to achieve a state where the corresponding atomic formula is a true belief. For example, `!picked_tomato(x,y)` states that the robot wants to pick the tomato at the given coordinates `(x,y)` in the playground. A test goal is denoted by the operator '?' followed by an atomic formula. It specifies that the agent wants to test if the corresponding atomic formula is true or not. For example, `?white_corn(x,y)` states that the robot wants to verify if the corn at the given coordinates is white or not.

Since the environment is dynamic, whenever the agent perceives a change in the environment, by means of its sensor, it could acquire new beliefs/goals or delete existing beliefs/goals. These four events are called *triggering events*, and they are denoted by prefixing with the '+' or '-' operator a particular belief/goal - e.g. sensing that a corn is black could be written as `+black_corn(600,972)` while aborting the goal of picking tomato is represented as `-!picked_tomato(x,y)`.

In AgentSpeak, every plan has a *head* and a *body*; the head consists of a triggering event, which causes the activation of the corresponding plan, and of a *context*[1], i.e. a sequence of atomic beliefs: to select a plan for execution, all the beliefs of the context must be a logical consequence of the beliefs in the knowledge base of the agent. In other words, all the terms of the context must evaluate as true. The body of a plan contains a sequence of actions and/or (sub)goals that the agent should execute, achieve or test to complete the plan.

For example, let us suppose that we want to write a plan that gets triggered when the robot detects a tomato. Before picking the tomato, we must check if the robot can actually carry it. Let us suppose that the robot can carry up to 6 tomatoes. Using the AgentSpeak notation, the plan is written as follows:

```
+tomato(X,Y) : caught_tomatoes(N) &
               (N < 6)
          <- reach_tomato(X,Y);
             +!picked_tomato(X,Y).
```

The triggering event of our plan, `+tomato(X,Y)`, is the *addition* of a belief whose terms represent the coordinates of the item. After the ':', the context is specified: in this example we check if, according to the current knowledge, the total number of caught tomatoes is actually less than six. Separated from the head by

---

[1] In our framework PROFETA we will use the term *condition* to refer to the same concept.

'<-', the body of our plan consists of two elements: first we have the action of reaching the tomato at the given coordinates, and then the achievement goal of picking it.

The declarative approach greatly simplifies the definitions of *strategies*, i.e. the set of plans which drive the agent towards the achievement of its goals. Modelling the behaviour of an AMR by using beliefs, goals, and plans is more simple and intuitive than using conventional imperative constructs. For these reasons, PROFETA applies the same principles of AgentSpeak to Python to provide a flexible tool for defining strategies.

3   The PROFETA Framework

PROFETA is a framework which allows the use of declarative constructs into Python programs. Since it has been written in Python, it exploits the object-oriented approach, the meta-programming and operator overloading abilities of this language to support AgentSpeak syntax and semantics.

PROFETA provides four basic classes, `Belief`, `Goal`, `Action` and `Engine`. The first two classes are used to map the corresponding concepts of the BDI model, the third class is an abstraction for all the actions a robot is able to perform, the latter class implements the engine which keeps track of the defined rules, is able to process them and to execute the associated actions.

The `Belief` class is meant to be used as base class for all custom beliefs. For example, say we want to create the belief `corn(150,722)`: first of all, we define the class `corn` as follows

```
class corn(Belief):
    pass
```

Then, we can create such a belief simply instantiating an object of that class providing, if needed, one or more parameters, for example "`corn(150,722)`". The polymorphism of the constructor of the base class `Belief` allows a programmer to pass any desired parameter(s), which will be stored in an internal attribute of class `Belief`, namely the `self._terms` list; the same subclass, in our example `corn`, can be thus used to represent many beliefs with the same meaning (provided that different parameters are used in the instantiation of it), say the position of the various corns in the playing arena.

The `Goal` class can be instead used as base class to define a custom *achievement goal*. According to PROFETA syntax, every goal has to be preceded by a '~', which is a syntactic replacement for the "!" symbol used in AgentSpeak. Hence, to create the goal `!collect_corns()`, we first define the class `collect_corns`

```
class collect_corns(Goal):
    pass
```

then we can create such a goal by invoking its constructor preceded by the '~' operator: `~collect_corns()`.

As in AgentSpeak, *triggering events* are denoted by using the '+' or the '−' operator before an instance of a belief or a goal, with the only exception that, in PROFETA, deletion of plans is not supported[2]. Hence, examples of legal triggering events are:

```
+corn(150,722)
-corn(150,722)
+~collect_corns()
```

Once we have had a look at `Belief` and `Goal`, which deal with the mental state of the robot, let us now examine the `Action` class. This class exposes the `execute()` method, which can be overridden in order to define concrete actions; as an example, let us write the code to represent the `reach_corn(150,722)` action, i.e. an action that moves the robot towards the specific point where a corn is located:

```
class reach_corn(Action):
    def execute():
        # get corn coordinates (given in the object's constructor)
        (X, Y) = (self._terms[0], self._terms[1])
        # reach the corn
        self.motor.forward_to_point ( (X, Y) )
```

To actually create the action, we will call its constructor and pass it the co-ordinates of the corn we want to reach, that is "`reach_corn(150,722)`"; as in `Belief`, any parameters provided during the instantiation are stored in the internal attribute `self._terms`.

Once we have defined a suitable set of subclasses of `Belief`, `Goal` and `Action` we are ready to define a plan. PROFETA allows to define plans using the "`|`" and "`>>`" operators; the syntax is the following

```
Head | Condition >> Body
```

Here, *Head* is a triggering event, *Condition* is a sequence of beliefs, separated by '&' which must be true for the plan to be triggered, and *Body* is a list that can contain both actions and triggering events. If the plan has no associated conditions and thus has to be activated simply following a certain event, the *Condition* part in the expression above can be omitted; in this case, plan definition is done as: `Head >> Body`. For example, if we want to write a plan that has to be executed when a corn in position (150, 722) is detected, we will write:

```
+corn(150, 722) >> [ reach_corn(150, 722), grab_corn() ]
```

---

[2] Plan deletion is not so useful, in practice, and its handling implies many issues that, from the semantic point of view, are not clearly pointed out in the definition of AgentSpeak [15,26]; for these reasons, in the current release of PROFETA, we decided to not support such a feature.

provided that `reach_corn` is the action defined above and `grab_corn` is another action that drives the corn picking mechanism of the robot.

It's now worth to describe the mechanism exploited by PROFETA to define plans. As reported in the beginning of this Section, the aim of PROFETA is to add declarative constructs to Python, which is instead an imperative object-oriented language. To this aim, the basic technique used is to exploit operator overloading and thus redefine the meaning of operators "+", "|", "~" and ">>". Indeed, from the Python point of view, the piece of code above defining the plan related to +corn(150, 722) event is a mathematical expression that, when found during execution, is *evaluated* by properly applying the unary "+" and right shift ">>" operators: by redefining their meaning in the operands, i.e. `Belief` and other classes of PROFETA, the various parts of the plan can be obtained and provided to the class `Engine`, which has the task of storing them and execute the associated actions when detects the occurrence of the associated triggering event.

In summary, Python checks the syntax and executes the *mathematical expression*, but does not know anything about its meaning, while the `Engine` class, together with the code provided in the methods redefining the operators, has the task of understanding the *semantics* of the plan definition, treating the latter accordingly. This is the reason why the *Body* of PROFETA cannot directly contain pieces of code (but only a list of specific objects), as they would be executed by the Python interpreter during expression evaluation; instead, the code relevant to actions to be done has to be encapsulated into `Action` objects so that they can be instantiated during plan expression evaluation (done by the Python interpreter), while the action encoded in the `execute()` method will be actually executed by the `Engine` when the plan is triggered.

For the same reason, when plans need variables that will be then bound to actual values on the basis of the content of the knowledge base, a proper syntactic trick need to be employed. Here the special function `_(..)` can be used to denote a *variable* within the scope of a plan, e.g. to denote the variable `X` we will write `_("X")`. An example will clarify this and other aspects. Say we want to write a plan that has to be executed every time the robot perceives the position of a corn (denoted by belief `corn(x,y)`), given that the number of picked corns (denoted by belief `picked_corn(n)`) is less than six, we have to write:

```
( +corn(_("X"),_("Y")) |
    ( picked_corn(_("N")) & (lambda : N < 6) ) ) >>
        [
          reach_corn(_("X"),_("Y")),
          grab_corn(),
          -picked_corn(_("N")),
          +picked_corn(_("N") + 1)
        ]
```

The plan above shows two additional syntactic elements that can be used to express a condition within the definition of a plan: the "|" and *lambda expressions*. The former can be used to separate the head of the rule (triggering event) from the body of the plan, while the latter allows a programmer to express boolean predicates that variables have to meet for the plan to be executed. Since the robot is able to store no more than six corn, the plan to execute when a corn is detected in a given position must be subject to the fact that the robot does not already possess six corns. Since the amount of corn picked is represented by the belief `picked_corn(_("N"))`, we must ensure that variable "N", bound to the `picked_corn` belief, is less than six. The use of the lambda expression[3] is mandatory to ensure (once again) that the predicate is computed by the `Engine` when the triggering event occurs, and not by the Python interpreter when it evaluates the expression representing the overall rule.

As introduced many times in this Section, core functionalities of PROFETA are implemented in the `Engine` class. It contains two main data structures: the *knowledge base* (KB)—the set of beliefs that represents the current robot's knowledge—and the *plan library*—the set of rules that are defined in the application according to the syntax reported above. The basic working scheme of `Engine` is to run a continuous loop that *(i)* checks if an event occurred, *(ii)* selects an appropriate plan, *(iii)* evaluates the condition verifying whether it is true, *(iv)* executes the plan[4]. While internal events (i.e. events specified in the Body of a plan) are directly handled by `Engine`, external events that are bound, for example, to perception made by sensors must be notified to the `Engine`, so that it can reason about them and determine the appropriate plan(s)—if any. In this case, to let the `Engine` know that an event has occurred, the `generate_external_event()` has to be used. For example, let us suppose the robot's sensors perceived the presence of a corn in the location whose coordinates are (150,722); to notify such a condition we need to obtain the instance of the `Engine` class[5] and invoke the `generate_external_event()` method, that is:

```
Engine.instance().generate_external_event( +corn(150,722) )
```

Since `Engine` holds the KB and the plan library, it exposes some methods to manipulate these structures. They can be used to print KB and plans, to obtain the beliefs that match a given belief pattern, to add or remove a belief, to activate or deactivate a plan, etc.

---

[3] A lambda expression in Python defines a *functor*, i.e. an function object that can be evaluated when needed.

[4] Indeed, the real working scheme of a BDI engine is quite more complex that described; due to space restrictions, we reported only a simplified version; interested readers may refer to [15,26].

[5] The `Engine` class is a *singleton*, i.e. there can exist only one instance of such a class.

4 Case Study

In this section we will show how PROFETA makes easy both to *write* and *edit* the behaviour of an AMR; a code snippet of the strategy for the scenario of this year's Eurobot competition is presented.

Let us consider a robot which is equipped with different devices to collect playing elements; in particular the robot has a ball sucker on the front side which makes it able to collect tomatoes by simply getting close to them and a harvester on the back side to collect ears of corn. Let us assume that, during the robot initialization, the knowledge base is filled with beliefs like corn(x,y) for every corn present on the playing table[6]; such a fact states that a corn is located at the specified coordinates but the robot doesn't know whether such a corn is actually black or white. Whenever sensors detect the color of a specific corn, the corresponding belief corn(x,y) is replaced with white_corn(x,y) or black_corn(x,y) accordingly. Prior to catch a corn, the robot has first to check its color: in order to design a robust strategy, we should write a set of plans that covers both of the situations that the robot could face. For instance, say we want to define a plan to collect four playing elements: two corns and two tomatoes. Such a plan could be written as follows:

```
+~pick_four_items() >>
    [
      +~pick_corn(150,722),
      reach_tomato(150,972),
      reach_tomato(600,1222),
      +~pick_corn(600,972)
    ]
```

The action reach_tomato(x,y) makes the robot move towards given coordinates to get closer to the tomato and thus grab it. The goal ~pick_corn(x,y) is associated with several plans, whose conditions represent the different situations the robot could have to meet. For example, let us consider the following plan:

```
+~pick_corn(_("X"),_("Y")) | corn(_("X"),_("Y")) >>
    [
      reach_corn(_("X"),_("Y")),
      detect_color(),
      +~pick_corn(_("X"),_("Y"))
    ]
```

This plan is selected when the target corn is actually present at the given coordinates, but its color is still unknown. In this case, the robot has first to approach the corn and detect the color: the knowledge will be updated by removing the corn(x,y) belief and adding either the white_corn(x,y) or the black_corn(x,y) belief. Eventually the event ~pick_corn(x,y) is re-triggered, and one of the two following plans is selected:

---

[6] The coordinates of all playing elements are known.

```
+~pick_corn(_("X"),_("Y")) | white_corn(_("X"),_("Y")) >>
   [
    reach_corn(_("X"),_("Y")),
    grab_corn()
   ]

+~pick_corn(_("X"),_("Y")) | black_corn(_("X"),_("Y")) >>
   [
     skip_corn(_("X"),_("Y"))
   ]
```

If the knowledge base states that the corn is white, the robot has only to reach it and then grab it. Otherwise, the robot has to skip collecting the black corn.

Now let us suppose that we want to use a different path to collect the items. To implement this new strategy, we have only to rewrite the previously defined plan as follows:

```
+~pick_four_items() >>
    [
     +~pick_corn(150,722),
     reach_tomato(150,972),
     +~pick_corn(600,972),
     reach_tomato(600,1222)
    ]
```

The robot will now execute a different sequence of actions, that is, a completely different *intention*; there is no need to modify other previously defined plans or pieces of code. Note that by using PROFETA we were able to specify the behaviour of our robot in a radically different way with respect to the use of traditional, less intuitive imperative procedures. In fact, we managed to define strategies using a language and a reasoning process whose operating principles recall the ones of the human mind.

5   Related Work

The literature on techniques and architectures to control an AMR is quite vast, as proved by the bibliography reported in [1]. And since AMR research is on a borderline between artificial intelligence and control theory, techniques proposed can be more AI- or control-related, on the basis of the skills of researchers. Many AI techniques have been successful employed in AMR programming, such as inference systems, neural networks or fuzzy logic, and the proposal of using an advanced reasoning system, like a BDI-based engine, is in accordance with such a line of research. Even if AgentSpeak provides some naïve examples of robot programming, to explain the syntax and semantics of the language, as far as we know, this paper is the first attempt to use a BDI-engine in a real autonomous robot. Some of the authors investigated the use of *rule-based inference engines* (i.e. expert systems) in [13,14,4], which have been successfully applied in the robots participating, for the UNICT-TEAM, to editions from 2007 to 2009 of the Eurobot competition. While expert systems, which are based on rules expressing the computation (action) to execute on the basis of a certain knowledge (beliefs), have proved their validity, the BDI model, as demonstrated in Section 4, provides abstractions and constructs that make strategy programming more similar to human reasoning.

Since AgentSpeak is a kernel language and cannot directly be used for implementation, some researchers have proposed a Java tool, called Jason [25], that implements the basic constructs and concepts of such a language. Jason is a tool providing a BDI reasoning engine and an interpreter of a Prolog-like language that has to be used to write the plans. Even if Jason has many similarities with PROFETA, its usage in an AMR presents some peculiar problems, often not so simple to solve. The first problem is practical: as it is widely known, the Java runtime environment is quite hungry of resources (CPU and memory), and thus it is hard to fit a small robot often equipped with an embedded system with a limited computing power. In this sense, the Python environment is quite light and performs quite well also in systems with few megabytes of memory[7]. The second problem is more conceptual and is related to the different languages a programmer has to use in Jason: here, the reasoning part has to be written in the Prolog-like language, while actions need to be coded in Java. PROFETA has been instead designed to offer a *all-in-one* environment, providing the same language and constructs to program both the reasoning/declarative and imperative part of a robot application, thus greatly simplifying its design, development and debug.

6   Conclusion

This paper has described the authors' experience in programming strategies for an autonomous mobile robot, designed for Eurobot 2010 competition, using a reasoning model called *Belief-Desire-Intention*. The paper has stated the characteristics of such a model and has presented a suitable implementation in a Python framework, called PROFETA, developed by authors. As it has been described in the paper, PROFETA has been successfully applied in the robot we built for the 2010 edition of Eurobot, and its advantages in writing and refactoring robot strategies have been proved.

Even if this is the first release of PROFETA and it performed quite well for our purposes, several issues need to be still dealt with and solved. Since a reasoning engine is always quite complex, from the computational point of view, evaluation of performances of PROFETA is a task that has to be done, in order to discover inefficiencies and apply proper optimizations. In addition, other features of AgentSpeak, not currently implemented in PROFETA, need

---

[7] Our target platform is a 160MHz ARM9-based system, with 64 MBytes of RAM and Debian Linux.

to be supported, as goal deletion or test goal. These and other issues will be dealt with in future work.

Acknowledgments. The authors would like to thank all the people of the UNICT-TEAM 2010 for their unvaluable support. Our deepest appreciation goes to Sebastiano Gennarini, Riccardo Massari, Andrea Milazzo, Rocco Milluzzo, and Daniele Tricoli for their precious contributions and advices.

References

1. Siegwart, R., Nourbakhsh, I.: Introduction to Autonomous Mobile Robots. MIT Press, Cambridge (2004)
2. Arkin, R.C.: Behaviour-based Robotics. MIT Press, Cambridge (1998)
3. Murphy, R.R.: Introduction to AI Robotics. MIT Press, Cambridge (2001)
4. Santoro, C.: An erlang framework for autonomous mobile robots. In: ERLANG 20007: Proceedings of the 2007 ACM SIGPLAN Workshop on Erlang. ACM Press, New York (2007)
5. Russell, S., Norvig, P.: Artificial Intelligence: A Modern Approach, 2nd edn. Prentice Hall, Englewood Cliffs (2003)
6. Forgy, C.: The OPS Languages: An Historical Overview. PC AI (September 1995)
7. JESS Home Page (2003), http://herzberg.ca.sandia.gov/jess
8. CLIPS Home Page (2003), http://www.ghg.net/clips/CLIPS.html
9. Drools Home Page (2004), http://www.drools.org
10. Di Stefano, A., Santoro, C.: eXAT: an Experimental Tool for Programming Multi-Agent Systems in Erlang. In: AI*IA/TABOO Joint Workshop on Objects and Agents (WOA 2003), Villasimius, CA, Italy, September 10-11 (2003)
11. Di Stefano, A., Santoro, C.: On the Use of Erlang as a Promising Language to Develop Agent Systems. In: AI*IA/TABOO Joint Workshop on Objects and Agents (WOA 2004), Turin, Italy, October 29-30 (2004)
12. Di Stefano, A., Santoro, C.: Designing Collaborative Agents with eXAT. In: ACEC 2004 Workshop at WETICE 2004, Modena, Italy, June 14-16 (2004)
13. Stefano, A.D., Gangemi, F., Santoro, C.: ERESYE: Artificial Intelligence in Erlang Programs. In: ERLANG 2005: Proceedings of the 2005 ACM SIGPLAN Workshop on Erlang, pp. 62–71. ACM Press, New York (2005)
14. Nicosia, V., Santoro, C.: Experiences from Using Erlang for Autonomous Robots. In: In Proc. of 12th International Erlang User Conference, EUC 2006, Stockholm, Sweden (2006)
15. Rao, A.: AgentSpeak (L): BDI agents speak out in a logical computable language. In: Perram, J., Van de Velde, W. (eds.) MAAMAW 1996. LNCS, vol. 1038, pp. 42–55. Springer, Heidelberg (1996)
16. Rao, A., Georgeff, M.: BDI agents: From theory to practice. In: Proceedings of the First International Conference on Multi-Agent Systems (ICMAS 1995), San Francisco, CA, pp. 312–319 (1995)
17. Bratman, M.E.: Intentions, Plans and Practical Reason. Harvard University Press, Cambridge (1987)
18. Wooldridge, M., Ciancarini, P.: Agent-Oriented Software Engineering: The State of the Art. In: Ciancarini, P., Wooldridge, M.J. (eds.) AOSE 2000. LNCS, vol. 1957, pp. 55–82. Springer, Heidelberg (2001)

19. Jennings, N.: On agent-based software engineering. Artificial Intelligence 117(2), 277–296 (2000)
20. Bradshaw, J.M. (ed.): Software Agents. AAAI Press/The MIT Press (1997)
21. Weiss, G. (ed.): Multiagent Systems. The MIT Press, Redmond (1999)
22. Pokahr, A., Braubach, L., Lamersdorf, W.: Jadex: A BDI reasoning engine. Multiagent Systems Artificial Societies and Simulated Organizations 15, 149 (2005)
23. Howden, N., Rönnquist, R., Hodgson, A., Lucas, A.: JACK intelligent agents-summary of an agent infrastructure. In: 5th International Conference on Autonomous Agents, Citeseer (2001)
24. Ingrand, F., Georgeff, M., Rao, A.: An architecture for real-time reasoning and system control. IEEE Expert 7(6), 34–44 (1992)
25. Jason Home Page (2004), http://www.jason.sourceforge.net/
26. Hübner, J.F., Bordini, R.H., Wooldridge, M.J.: Programming declarative goals using plan patterns. In: Baldoni, M., Endriss, U. (eds.) DALT 2006. LNCS (LNAI), vol. 4327, pp. 123–140. Springer, Heidelberg (2006)

Using Quadtrees for Realtime Pathfinding in Indoor Environments

Julian Hirt, Dominik Gauggel, Jens Hensler,
Michael Blaich, and Oliver Bittel

University of Applied Sciences Konstanz, Germany
Laboratory for Mobile Robots
Brauneggerstr. 55 D-78462 Konstanz
{jhirt,gauggel,jhensler,mblaich,bittel}@htwg-konstanz.de
http://www.robotik.in.htwg-konstanz.de/

Abstract. During the last few years mobile robots got more and more important to solve different tasks in outdoor and indoor environments. To solve these tasks one very essential issue is to get from one point A to another point B as fast as possible. To find the least expensive route to the goal the pathfinding process needs a full state space information about the environment. With this information we can use optimally efficient algorithms like A* to find the route, but this might be very expensive on memory usage and time response. therefore we need to use other data structures to represent the whole information about the environment. This paper shows how the usage of quadtrees improves performance in terms of computation speed, memory requirements and path length.

1   Introduction

The Eurobot Challenge [1] is an international open mobile robotics contest which is taking place every year. The task which the amateur teams have to handle in 2010s challenge is to build a robot which collects virtual food on a 2m x 3m board. The robot has 90 seconds to collect as much food as possible while another robot does the same on the same board. All the robots have to do so completely autonomous. therefore we needed an algorithm which finds the best path on the board between two points.

Pathfinding is a fundamental problem for mobile robots. No mobile robot could work on almost any task without moving to another point in the environment. therefore different algorithms exists which find good solutions to get from one point A to another point B. This problem has been defined as the problem finding a path linking two vertices of a graph [2]. There are currently different algorithms to solve the pathfinding problem like the Wavefront or A* algorithms. One of the most common solutions is to implement the A* search algorithm [3]. This algorithm has been first described in 1968 and has been used in many different ways. A* is optimally efficient for a certain heuristic. Yet, in practice, pathfinding using A* might have huge problem with memory and

D. Obdržálek and A. Gottscheber (Eds.): EUROBOT 2010, CCIS 156, pp. 72–78, 2011.

time requirements. This affects a mobile robot especially when routes have to be calculated very frequently.

The navigation process for mobile robots can be separated into two levels - global and local planning. The local planning is responsible for avoiding obstacles [4], reacting to sensory data and driving towards a subgoal. This paper focuses on combining the local and global navigation by using one map for both types.

The literature gives many different approaches to overcome the limitations of the A* search algorithm. Frequent advices are changing the heuristic, like Euclidean Distance or Manhattan Distance, caching parts of calculated routes or especially changing the search space representation. The most promising aspect is to change the underlying representation of the environment which is usually a grid [5]. The complexity of the pathfinding process expands with the size of the environment and their accuracy. For a higher accuracy more cells, which represents a certain area, are needed. By discretizing the world, the computational complexity of pathplanning can be controlled by adjusting the cell sizes. By enlarging the cells more space, which might not contain any obstacles or impassable regions, gets lost for the mobile robot. What we need is a data structure which represents the environment as accurate as possible with the least amount of nodes. So data structures with uniform cell sizes are not practical for this situation. Efficiency in map representation can be obtained by the use of quadtrees. In this paper we discuss our approach to solve the complexity of the pathfinding process using the A* search algorithm by using quadtrees to represent the environment of a mobile robot. We present our results of a path planner used for the Eurobot contest.

2   Map Representation

The discretization of a world offers the possibility to control the complexity of pathfinding algorithms as well as a flexible representation of regions which contains obstacles or just impassable regions. The simplest representation of a 2D world is a regular grid of squares. They support random-access lookup for any coordinates of a point in a map. To represent a world in a grid accurately the size of cells have to be as small as possible. On the one hand we need this accuracy to guarantee the quality of the solution path, but on the other hand the number of cells boost the complexity of memory and time requirements.

2.1   Regular Grid

Using a regular grid means to divide the entire map into uniform sized squares. Every square has four or eight neighbors, by adding the diagonal squares as well. This means if a region consists of a certain number of squares but shares the same information, it is still represented by the amount of squares. This represents the world inefficiently because it makes the search process very expensive. Moreover,

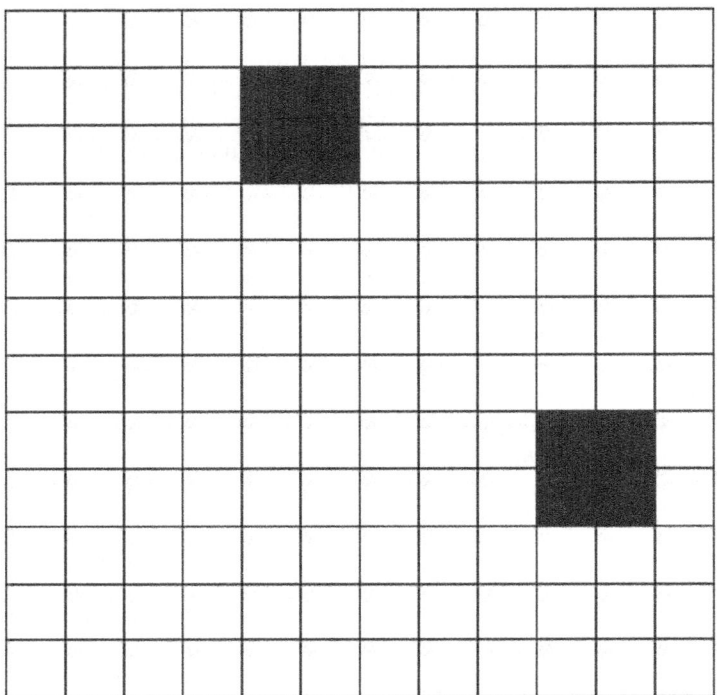

Fig. 1. A regular grid for a map which contains two obstacles. It consists of 144 uniform cells

a regular grid allows only eight angles which results in sudden direction changes. It might be possible to straighten a path through a clear area but there is no guarantee that this smooth path is the optimal path. How a map can be presented in a regular grid is demonstrated in figure 1.

2.2   Quadtree

To reduce the amount of squares to represent a world, we use a quadtree, which is a different data structure to describe the map. Compared to a regular grid a quadtree has no uniform sized cells. It subdivides the whole map into four equally sized regions. If one of these new regions contains an obstacle it is subdivided into four other regions as well. A region is recursively subdivided until a region does not contain any obstacle or reaches a certain minimum size. Quadtrees allow to represent a large clear area with one single cell. Instead of the regular grid it describes a world efficiently but at the cost of quality. Paths generated by quadtrees are suboptimal because they are constrained to use the centers of the cells as shown in figure 2.

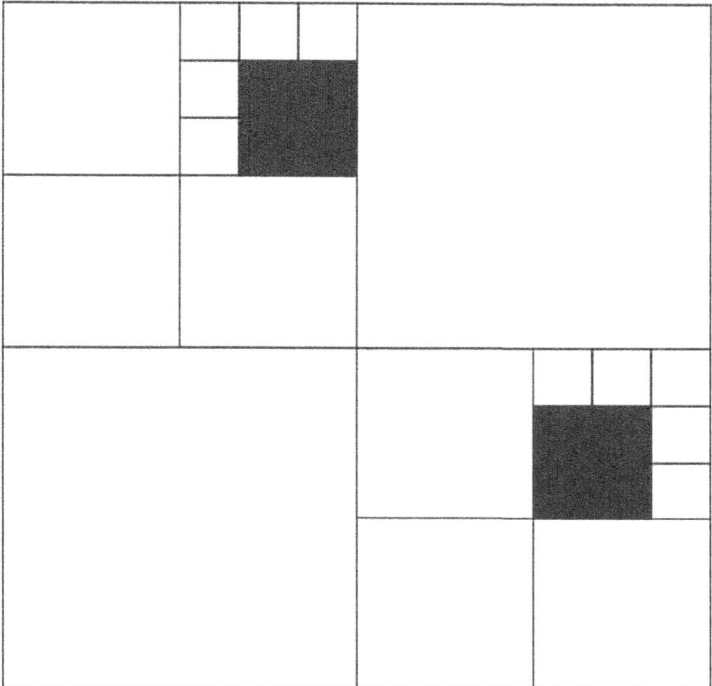

Fig. 2. A quadtree for a map which contains two obstacles. It consists of 26 cells

3   Pathfinding in a Quadtree

In our first approach we implemented a regular A* search algorithm. With a resolution of 1cm per pixel we represent our board by a regular grid with 60000 cells. Using this regular grid with an A* algorithm takes 0.75 seconds to calculate a path from one corner to the opposite corner. During a match of 90 seconds we have to drive to around 30 goals. Hence, 0.75s * 30 takes too much time to just calculate the paths. Especially because the board is very dynamically. Obstacles on the board can move and particularly the other robot might be a problem. Our robot is driving with up to 1m/s. At this speed it is too dangerous to drive 0.75s while calculating a new path. therefore we implemented a quadtree in our second solution to solve this problem.

To use a quadtree in our pathfinding process, we have to generate a quadtree out of the regular grid every time we need to run the A* algorithm. It subdivides the world into subregions until a region does not contain any other obstacles. This reduced our amount of nodes enormously from 60000 down to 990. An example for a calculated path between two corners is shown in 3. therefore we tested the system as a Player [6] driver on our robot. Each of the green squares represents one node in our quadtree. The violet lines show all squares which were visited by the A* search algorithm. The actual path which was found by the A* is the red line in figure 3. In order to use the quadtree for the local navigation as well, we need to update our grid every time the map changes. Thus we have to regenerate the quadtree to get a valid path for new goals.

3.1   Path Relaxation

As you can see the violet and red lines are from one center of a square to another center. This might give us a suboptimal path which we optimized by using the Split and Merge algorithm [7]. For the merge part of this algorithm we have to have a look at the centers of three following squares (s1,s2,s3) which are used in our path. If dist(s1,s3) is smaller than dist(s1/s2)+dist(s2/s3) we can merge them if there is no obstacle on the way from s1/s3. This smoothens our path and gives us a better path. This optimized path is shown in figure 3 as the cyan line.

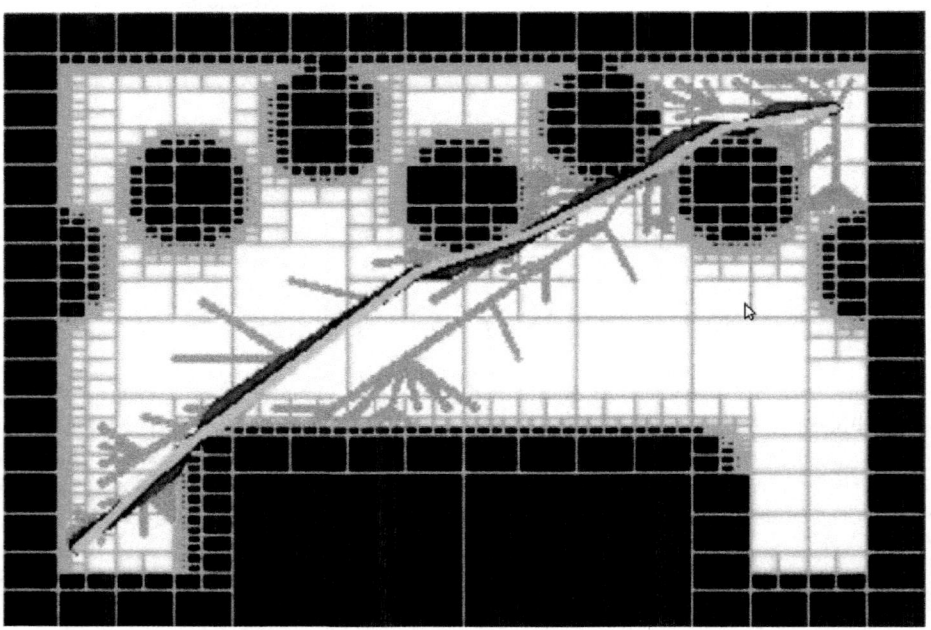

Fig. 3. Quadtree for calculated path between corners

4   Experimental Results

For the following results we used an Intel mainboard with an Atom 330 processor (2x1.6GHz) and 2Gb memory. Instead of just running the A* search algorithm on a regular grid, we need three different steps using a quadtree. The breakup of the computation times is shown in table 1. The first step is to generate the actual quadtree, which takes 97.5% of the process time. To find a path with an A* algorithm needs only 2% which is about 0.0004s. This shows us the very enormous improvement compared to a regular grid which needed 0.75s. To simplify the resulting path takes another 0.0001s which is only 0.5% of the whole process time. Compared to the regular grid the quadtree needs only 2.73% of the time what we need to calculate the same path on a regular grid although we have to do two more steps.

Table 1. Breakdown of computation times using a quadtree

Part	Time used	%
Generate Quadtree	0.02s	97.5 %
A*	0.0004s	2.0 %
Split And Merge	0.0001s	0.5 %

Table 2. Comparison of the computation speed between A* on a regular grid and a quadtree

Type of Representation	Time used	%
Regular Grid	0.75s	100 %
Quadtree	0.0205s	2.73 %

Table 3. Comparison of the memory requirements between A* on a regular grid and a quadtree

Type of Representation	Amount of nodes	%
Regular Grid	60000	100 %
Quadtree	990	1.65 %

5 Conclusion and Future Work

We have implemented a method for a pathfinding process on a 2m x 3m board for a mobile robot. Our solutions combines the optimally efficient A* algorithm with the quadtree. This gives us the possibility to find an almost optimal path between two points in much shorter time than an A* in combination with a regular grid. The quadtree data structure optimizes the representation of our map by partitioning the map into non uniform sized cells. Although this might not result in an optimal path, the savings in time and memory requirements are way more significant. Even the time to generate a new quadtree does not affect that.

Actually, we might be able to improve the efficiency of our algorithm in our future work. It is not necessary to generate the whole new quadtree after the map changed. It is possible to generate just the regions which were changed. This might give us another huge improvement in the computation speed.

Another point would be to have a look at the framed quadtree [8]. We might be able to replace our Split and Merge algorithm by implementing a framed quadtree. With this technique we might be able to smoothen our resulting path even more.

References

1. Sciences, E.P.: (February 27, 2010), http://www.eurobot.org
2. Niewiadomski, R., Amaral, J.N., Holte, R.C.: A performance study of data layout techniques for improving data locality in refinement-based pathfinding. The ACM Journal of Experimental Algorithmics, 1–28 (2004)
3. Hart, P.E., Nilsson, N.J., Raphael, B.: A formal basis for the heuristic determination of minimum cost paths. IEEE Transactions on Systems Science and Cybernetics, 100–107 (1968)
4. Kunchev, V., Jain, L., Ivancevic, V., Finn, A.: Path Planning and Obstacle Avoidance for Autonomous Mobile Robots: A Review. In: Gabrys, B., Howlett, R.J., Jain, L.C. (eds.) KES 2006. LNCS (LNAI), vol. 4252, pp. 537–544. Springer, Heidelberg (2006)
5. Yap, P.: Grid-Based Path-Finding. In: Cohen, R., Spencer, B. (eds.) Canadian AI 2002. LNCS (LNAI), vol. 2338, pp. 44–55. Springer, Heidelberg (2002)
6. Collett, T.H.J., MacDonald, B.A., Gerkey, B.: Player 2.0: Toward a practical robot programming framework. In: Australasian Conference on Robotics and Automation, Sydney (2005)
7. Thorpe, C.E.: Path relaxation: Path planning for a mobile robot. In: AAAI 1984 Proceedings (1984)
8. Yahja, A., Stentz, A., Singh, S., Brumitt, B.L.: Framed-quadtree path planning for mobile robots operating in sparse environments. In: IEEE Conference on Robotics and Automation (ICRA), Leuven, Belgium (1998)

Robot Workshop and Contest for High-School Students Organized by "Politehnica" University of Bucharest

Sanda Paturca[1], Catalina Enescu[2], Constantin Ilas[1], and Alexandru Morega[1]

[1] "Politehnica" University of Bucharest, Faculty of Electrical Engineering,
Splaiul Independentei 313, 69121 Bucharest, Romania
sanda.paturca@upb.ro, constantin.ilas@upb.ro
[2] "Miguel de Cervantes" High School, Calea Plevnei 38-40, Bucharest, Romania

Abstract. The paper presents a workshop and a contest organized for 25 high school students at the "Politehnica" University of Bucharest in early 2010. These activities were coordinated by both University professors and high school teachers and their objectives include introducing students to robotics and also creating a bridge between engineering university and high school, to attract more high schools students towards engineering education. The events showed that the students were very motivated when working together to solve practical problems and when they were in competition with their peers. Based on the success of this project we are considering extending it in the future, to cover several high schools.

Keywords: robot workshop, robot contest, high-school students, engineering education, high-school curricula.

1 Introduction

This paper presents a workshop and a contest organized for 25 high school students at the "Politehnica" University of Bucharest in early 2010. The students were in their 11th grade, corresponding to an age of 16-17, at the "Miguel de Cervantes" High School in Bucharest. The organizers were university professors that also run the application lab for the Robotics discipline at the Faculty of Electrical Engineering, together with their high school teacher of Informatics and Communication Technology.

The main objectives for organizing the workshop and the contest were:

- Introducing, involving and get interested the high school students in solving a technical problem and introducing them to engineering. As in many other countries, engineering is not always on the top of the list of potential disciplines for high-school graduates. Through the contacts with university professors and students we wanted them to become more familiar with engineering and help them to be fully informed when choosing a university specialization.

- Creating a pilot, based on which we could extend this project to cover people from several high-schools.

- Introducing the high school students to robotics and engaging them in solving a practical problem using the friendly LEGO Mindstorms NXT kit.

D. Obdržálek and A. Gottscheber (Eds.): EUROBOT 2010, CCIS 156, pp. 79–86, 2011.

The workshop took place in four sessions of 2 hours each, over a period of 3 weeks, and the contest was organized immediately after.

In the next sections we will detail the organization of the workshop and the contest, the feed-back and the lessons learnt, and also the revealed ideas for future similar projects.

2   Previous Experience

Some of the professors had already worked together with the high school teacher before, during a couple of research and education projects. For example, one of these educational projects, based on advanced Dynamic C programming for industrial applications, was organized between 2005 and 2007 and it's final goal was to prepare high school and university students in Computer Science and Automation field.

Also, the Faculty of Electrical Engineering runs a class of Robotics, which includes an application lab. At this lab students develop applications of robot navigation, obstacle avoidance, manipulators and they participate at 1-2 competitions during the class. These activities showed us the benefits of the competitions and gave us some experience on how to organize them.

Furthermore, parts of the platforms we use during this lab include a set of Mindstorms NXT robots from Lego with whom we were able to develop several interesting applications, despite of their limitations. Also we were very aware that such platforms are easy to work with and can be programmed at a high level (in NXC, a language very close to C), being ideal for beginners with some knowledge on programming.

3   Workshop Description

The workshop was organized in four sessions of 2 hours each, over a period of 3 weeks. The workshop was created to be suitable for beginners, with the objective of encouraging them to enter into the world of robotics. The main purpose was to help them become familiar with the Mindstorms NXT robot kit and also use the freshly acquired knowledge to develop their application for the contest.

The first practical session consisted mostly in a description that we provided on the robot kit, its main components (actuators, sensors, control unit) and its programming, as well as the first simple applications they developed. We presented them several programming options but we insisted on programming in Not Exactly C (NXC). This language is very close to ANSI C (as its name suggests) and we selected it because the students were already familiar with ANSI C and because it provides a powerful way of developing applications for this robot.

The NXT robot features a simple Operating System (OS), therefore it is possible to support a form of multi-threading, in which multiple tasks can be run independently on the same resources, based on a time-division and event-driven strategy [1], [2]. However, we decided not to insist on the concept of multi-threading and its possible implementations on this robot using the basic OS included in the firmware. The

Fig. 1. One of the teams that participated at the workshop and the contest

reasons for this decision were the fact that students were not familiar with this concept and because for simple applications such as those used in the workshop as well as the one they developed for the contest, multi-threading was not necessarily needed.

At the end of this session the students were able to create very simple programs, compile, download and execute them on the NXT robot.

At the beginning of the second session we introduced the contest, to be held after the workshop completion and presented them the contest topic and rules. The topic consisted in creating and programming a robot that would receive commands via the microphone and would be able to start (at command) and go straight, turn around at 180 degrees (at command), then go straight (in the reverse direction) and stop (at command). Full details on the topic and the rules are presented in section 4 below. From this point forward, the participants worked in teams and the workshop was focused on gradually helping them understand the basic concepts that they needed in order to complete this task and letting them gradually build their application. One of the teams that participated in the workshop and in the contest is presented in Fig. 1.

Since the contest was focused on robot motion, we started with a brief presentation on the basics of motors and position control of motors. We only presented the DC motor, as it is easier understood by high-school students. Regarding the position control, the students learned about the closed loop principle and the related elements – position sensor feedback and control tasks that are usually performed by the robot.

During this session the challenge was to keep the students' attention and motivation while presenting these concepts that were new to them. We tried to make this part as interactive as possible, thus ensuring that the pace and the level of details are well linked to their learning abilities.

We let them build a first version of the robot and then program it to go straight and turn around at 180 degrees. One of the robots they developed at that stage and that was later used in the contest is presented in Fig. 2. We recommended them to use,

Fig. 2. One of the robots developed during the workshop for the contest

only for this session, hard coded start and turning moments, and to focus on achieving the straight movement and the turning. We chose to let them figure out how different turning strategies can be implemented and not present them in advance. Practically, we needed to have some individual coaching dialogues with each team to help them figure out and then implement the turning movement.

During the third session we presented to students the sensors included in the NXT kit: ultrasound, touch, sound and light sensors, and the students were encouraged to test them. At this point we discussed about several possible ways for a robot to get a command during its movement and we decided together with the students that we should use a sound type of command and consequently a sound sensor (microphone). Practically the students easily deduced that the simplest way to solve their problem was to use the same command (e.g. clapping the hands) that could mean different actions depending on the current state of the robot or the command position during the chain of commands received (start for the first command, turn around for the second, stop for the third). The students easily understood the concept of having a threshold to distinguish between command and background noise and enjoyed finding alternative ways of generating the sounds (hitting the floor with a foot, hitting a metal object, etc). They also understood how the robot reacts to the sound of a given command, taking into account that the sound sensor is read continuously in a loop. They discovered that the robot firstly didn't make anything at command because it practically made the whole chain of tasks, including stop in a single sound command. They understood that it is needed to insert a minimum value for the time between two successive actions, which must be greater than the duration in which the sound intensity of one command remains over the threshold level.

On the forth session the students worked on finalizing the correct robot behavior related to the command and on refining their solution. They worked completely independent but we had a couple of short discussion with each team to track their progress and answer their questions. The challenges that they faced included finding a threshold that would not enable false commands but also not cut the correct ones, turning at exactly 180 degrees and especially having the robot move in perfect straight line. They found relatively easily the correct solutions to the first two issues. For solving the first issue the students wrote a test program to record the sound level of their input commands and visualized, on the PC, the sound waveform. For the third issue they tried several approaches. Because they had no knowledge of control theory they did not have the idea of measuring the distance traveled by each of the two motor wheels and compensate the motors power by the difference between these two distances. However, most of them managed to have satisfactory results. All teams also realized that in order to win they had to cover the track as fast as possible, focusing on obtaining a reasonable precision in track following.

4 Contest Description

The robotics competition was held at the end of the robotics course in order to offer them the opportunity to put in practice what they learned. They had to work in teams to design and develop a program to solve the given problem, test it and run it in the contest. Prizes were awarded to the best performing teams.

The contest topic consisted in creating and programming a robot that would receive commands via the microphone and would be able to start (at command) and go straight, turn around at 180 degrees (at command), then go straight (in the reverse direction) and stop (at command). We showed them the basic configuration of the

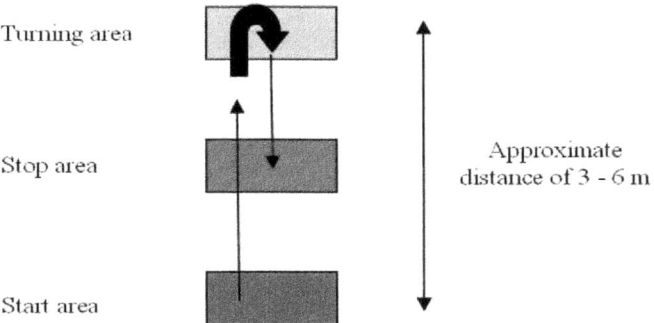

Fig. 3. Configuration of the contest circuit, as introduced to participants during the workshop. Exact distances were subject to announcement in the contest day.

circuit, as presented in Fig. 3. At that moment we gave them only the approximate length of each movement segment, the exact lengths being subject of announcement in the contest day.

The time from start to stop was measured. The robot had to go as straight as possible and turn so that it stayed within the turning area during turning, then go straight and stop within the stop area. If the robot went outside the turning area during turning or it stopped outside the stop area, a penalty of 10% of the total time was applied. The team that was able to go through the circuit during the shortest time (including possible penalties) won. There were 6 teams, 5 of them consisting of 4 people and one of 5 people. During the contest day it was presented the exact track. Each team was allowed to make a couple of tests, to get familiar with the track and to do last minute tuning.

Table 1. Summary of the results for each team during the two rounds. Best result out of two tries was recorded

Round no.	Result	Team no. 1	Team no. 2	Team no. 3	Team no. 4	Team no. 5	Team no. 6
1	Time (s)	17.5	21.5	19.2	18.0	15.2	17.6
	Penalty	10%	10%	10%	-	10%	10%
	Total (s)	19.3	23.6	21.1	18.0	16.7	19.4
2	Time (s)	16.3	DNF	20.2	18.8	16.1	15.9
	Penalty	10%		10%	-	-	10%
	Total (s)	17.9		22.2	18.8	16.1	17.5
	Rank	3	6	5	4	1	2

The actual contest consisted in two rounds and the teams had two tries in each round. The best result out of the two tries was taken into account. The aim of the first round was just to determine the starting order for the second round – and indirectly to give them a further chance to do any needed tuning. The final results were those recorded in the second round. The results are summarized in Table 1.

As can be seen three teams managed to improve their result in the second round, while two were slightly slower and one did not manage to finish. The total track length was 9 m, so considering that their turn took around 1 - 2 s, the average speed was around 2 km/h.

After the contest all students received a workshop participation certificate and the first three teams during the contest were awarded prizes.

Some pictures from the contest and from the award ceremony are presented in Fig. 4 below.

Fig. 4. Pictures from the contest (a), (b), (c) and from the award ceremony (d)

5 Future Activities

Based on the success of this project, we intend to continue our efforts of involving high school students into engineering activities. Our intentions are to provide part of this workshop on an e-learning platform, thus enabling students to work at their own pace and select the experiments and make this activity accessible also to students from other areas. We also intend to help high schools to create High School Robotics clubs, where various basic engineering concepts can be covered. These Robotics clubs may eventually evolve to an optional Robotics class, in the curricula. Finally, we intend to expand the competition by having a local phase, within several high schools and then the best teams from each high school meet into a final, organized by the University.

6 Conclusions

The workshop and the contest proved to be an excellent opportunity for the high school students to learn about robotic concepts and apply this by hands-on activities.

The students solved an engineering design problem in a competitive way. They were all motivated and they expressed their wish to have future similar competitions. As we could observe during the workshop and their opinion expressed in the feedback forms that we distributed after the workshop and the contest, the vast majority of students had fun, enjoyed the challenges and said they learned many new concepts. Students liked working with robots and gradually they wanted to stay in the classroom longer, in order to learn new robot hardware programming commands, test their design and make new experiments. They established positive and productive interaction among themselves by working together and through the brainstorming that this activity required. Participants think that this experience will help them to learn Math, Computer Science and become more familiar with basic engineering concepts. Majority of students said that they had great fun and now have a better understanding of Computer Science.

The team has been building a lot of self-confidence as a result of this experience. More importantly, they expressed their willing to continue to use in future activities what they have learned from the workshop.

The participants also strongly agreed that the partnership between the high-school and university is important. Everybody greatly benefited from this experience.

References

1. LEGO® Mindstorms® NXT® Direct commands LEGO Group (2006), http://mindstorms.lego.com
2. Astolfo, D., Ferrari, M., Ferrari, G.: Building Robots with Lego Mindstorms NXT, Syngress. Elsevier, Amsterdam (2007)
3. Balch, T., et al.: Designing Personal Robots for Education: Hardware, Software and Curriculum. IEEE Journal of Pervasive Computing 7(2), 5–9 (2008)
4. Siegwart, R., Nourbakhsh, I.: Introduction to Autonomous Mobile Robots. MIT Press, Cambridge (2004)
5. Wang, E.L., Lacombe, J., Rogers, C.: Using LEGO® Bricks to Conduct Engineering Experiments. In: Proceedings of the ASEE Annual Conference and Exhibition, Session 2756 (2004)
6. Cardeira, C., da Costa, J.S.: A Low Cost Mobile Robot for Engineering Education. In: IEEE IECON Conference (2005)
7. Perova, N., Rogers, C., Feldman, D.H.: Investigation of the successful effort to change educational curriculum frameworks in Massachusetts to include engineering and technology. In: American Society for Enginering Education Annual Conference, Austin (2009)
8. Danahy, E., Goswamy, A., Rogers, C.: Future of robotics education: The design and creation of interactive notebooks for teaching robotic concepts. In: IEEE International Conference on Technologies for Practical Robotic Applications, Woburn (2008)
9. Perova, N., Johnson, W., Rogers, C.: Using LEGO-based engineering activities to improve understanding concepts of speed, velocity, and acceleration. In: American Society for Engineering Education Annual Conference & Exposition, Pittsburgh (2008)

Requirements and Solutions in Applied Robotics

Rahman Jamal

National Instruments Central Europe, Ganghoferstr. 70b
80339 Munich, Germany
rahman.jamal@ni.com

Abstract. Robotics is one of the fastest growing engineering fields, and one of the most challenging. When designing, prototyping, and deploying robotics applications, three of the biggest challenges are integrating with sensors and actuators, implementing autonomy, and deploying deterministic control algorithms to embedded hardware. This paper introduces LabVIEW Robotics, a new graphical software framework for designing sophisticated autonomous systems. It offers developers the ability to use one software development environment for designing control algorithms, connecting to real-world I/O, and deploying to deterministic hardware targets.

Keywords: robotics applications, graphical programming, prototyping, deployment, deterministic control, embedded hardware, autonomous vehicles.

1 Introduction

Robots are becoming a part of everyday life. According to IFR World Robotics, today approximately 8.6 million robots are around the world. They are mowing lawns, vacuuming living room floors, assembling hybrid cars, autonomously performing deep-water drilling or even challenging us with humanlike behavior. They serve the government, defense, medical, agricultural, mining, space, and many other industries by performing the tasks that are dull, dirty, or dangerous to humans. Undoubtedly, robotics has become one of the fastest growing engineering fields and at the same time one of the most challenging. Almost every type of robot operates in a different environment, has a different behavior or task, and connects to different sensors and actuators. Therefore, they are often developed on different hardware platforms with different software development tools. When an engineer develops a proven control system for one robot, it is difficult to transfer it to another robot because the APIs for sensing, autonomy, and motor control are different in syntax.

When designing, prototyping, and deploying robotics applications, three of the biggest challenges are integrating with sensors and actuators, implementing autonomy, and deploying deterministic control algorithms to embedded hardware. To address these challenges, LabVIEW Robotics provides an entirely new suite of robotics-specific sensor and actuator drivers and new libraries for complex navigational algorithms. In addition, LabVIEW offers developers the ability to use one software development environment for designing control algorithms, connecting to real-world I/O, and deploying to deterministic hardware targets.

D. Obdržálek and A. Gottscheber (Eds.): EUROBOT 2010, CCIS 156, pp. 87–92, 2011.

LabVIEW Robotics is built on the LabVIEW programming language, taking advantage of more than 20 years of proven development and unmatched productivity. LabVIEW Robotics can be subdivided into the following five areas:

- Intellectual property (IP) – New algorithms are specifically designed for creating complex, mobile systems

- Graphical programming – The fundamental building block of the LabVIEW programming language provides unmatched productivity for designing sophisticated mechatronics systems

- Deployment to real-time and FPGA hardware devices – Seamless embedded real-time and field-programmable gate array (FPGA) hardware integration makes implementing your robotics applications significantly easier than with traditional tools

- Connectivity to sensors and actuators – From standard robotics sensors such as ultrasonic and infrared (IR) sensors to higher-end devices such as light detection and ranging (LIDAR) sensors and controller area network (CAN)-based smart motors, LabVIEW Robotics comes complete with sensor and actuator drivers for Windows, real-time, and FPGA devices

- Easy integration of text-based tools – LabVIEW Robotics makes it simple to integrate existing code with the LabVIEW MathScript RT Module, as well as native C and HDL import capabilities

1.1 Intellectual Property (IP)

LabVIEW Robotics features a new Robotics palette containing algorithms for designing your next robotics controller. You can find everything from sensor drivers to inverse kinematics on the new Functions palette:

- Connectivity VIs for third-party software integration
- Obstacle Avoidance VIs for mobile systems
- Path Planning VIs to calculate a path to a goal point within a map
- Protocol VIs to process data formatted in communication protocols such as National Marine Electronics Association (NMEA) and Joint Architecture for Unmanned Systems (JAUS)
- Robotic Arm VIs to make dynamic and kinematic calculations on a robotic arm
- Sensing VIs that configure, control, and retrieve data from robotics sensors

1.2 Graphical Programming

The dataflow nature of LabVIEW is well-suited for designing robotics applications because this approach almost precisely mirrors the design process of a simple or complex autonomous system. For example, if your initial design is based on a "sense-think-act" architecture, this is clearly represented in LabVIEW, as shown in Figure 1. This architecture not only makes your design more straightforward but also helps your colleagues better understand your program through the intuitive nature of Lab-VIEW.

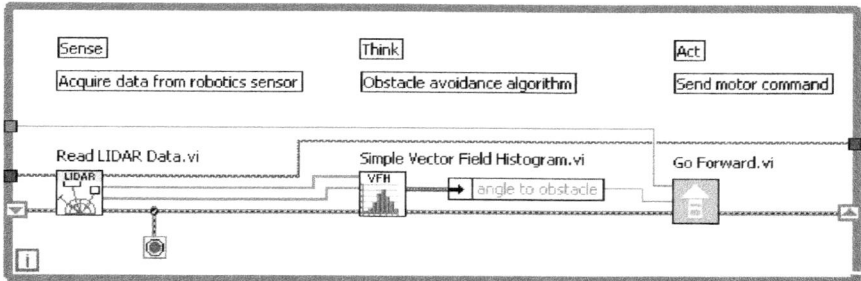

Fig. 1. The dataflow programming language of LabVIEW is well-suited for designing robotics applications through a "sense-think-act" architecture

1.3 Connectivity to Sensors and Actuators Deployment to Real-Time and FPGA Hardware Devices

A large gap in many robotics software tools available today is the inability to deploy your application to real-time and embedded hardware devices. Many tools allow you to design and explore your application on a Windows, Linux, or Macintosh platform but leave it up to you to get that same code onto an embedded platform. Lab-VIEW Robotics IP is specifically designed and optimized to easily deploy to NI reconfigurable I/O (RIO) targets including NI CompactRIO and NI Single-Board RIO devices.

1.4 Deployment to Real-Time and FPGA Hardware Devices

From its first release, the LabVIEW programming environment has helped you save time by connecting to instruments and devices. LabVIEW today is the de facto standard for instrumentation connectivity through the high quality and breadth of instrument drivers. LabVIEW Robotics takes advantage of this strength by incorporating an entire suite of robotics sensor and actuator connectivity options. These drivers remove the time-consuming task of writing, testing, and implementing sensor drivers for your robotics system. In fact, LabVIEW Robotics includes multiple versions of many sensor drivers for Windows, real-time, and FPGA-based platforms, ensuring you can connect your sensor appropriately for your I/O needs.

LabVIEW Robotics IP is specifically designed and optimized to easily deploy to NI reconfigurable I/O (RIO) targets including NI CompactRIO and NI Single-Board RIO devices.

1.5 Easy Integration of Text-Based Tools

The LabVIEW environment is best known for its graphical programming nature, but it also includes a breadth of integrated and imported utilities for text-based algorithms. You can easily incorporate your existing C, .m file, or HDL code into your LabVIEW Robotics application, or you can develop new algorithms within LabVIEW in C (Formula Node), .m files (LabVIEW MathScript RT Module), or HDL [HDL Node or component-level IP (CLIP Node) in the LabVIEW FPGA Module]. These options provide a variety of tools to ensure maximum reuse and offer the right programming model for the problem you are trying to solve.

2 Examples of Real-Worl Robotics Applications

LabVIEW Robotics provides many helpful example programs that demonstrate how to connect your sensor data and robotics IP to create a sophisticated, autonomous system. These examples show real-world use cases of assembling the individual robotics functions to perform a useful task. For example, you can use vision algorithms in several different application areas; the development team took the applicable functions and created new robotics examples that help you understand how to apply them to your autonomous applications for target tracking or path following. Other helpful examples include the following:

- Architectures – Helpful templates for simple and sophisticated robot design controllers
- Basics – Simple FPGA basic code for serial, SPI, and PMW connectivity
- Third-party connectivity – From simulators to Microsoft Robotics Developer Studio, Skilligent, MobileRobots, and other robot platforms
- Communication protocols – NMEA, SPI, RS232, and I^2C communication examples
- Control and simulation – DC motor control, PID, predictive observer, and Extended Kalman filter examples
- Motion control – PWM, CAN-based, and NI-Motion control
- Path planning – A*, AD*, and Voronoi examples
- Robotic arm – Serial robot, forward, and inverse kinematic examples
- Steering – Mecanum, omni, differential, and other steering examples
- Vision – Color tracking, path following, and target tracking

Some of the examples are actually project architectures, which serve as starting points for various robot applications.

The single control loop robot architecture in the Robotics Project Wizard serves as a starting point for robots that perform simple, repetitive algorithms. You can insert code for acquiring sensor data, processing data, and controlling the robot inside the timed loop.

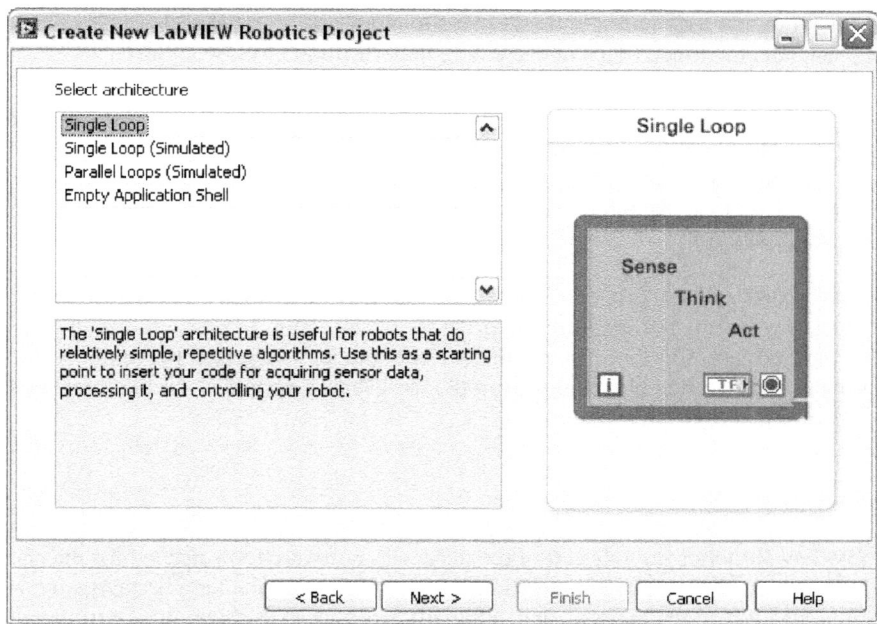

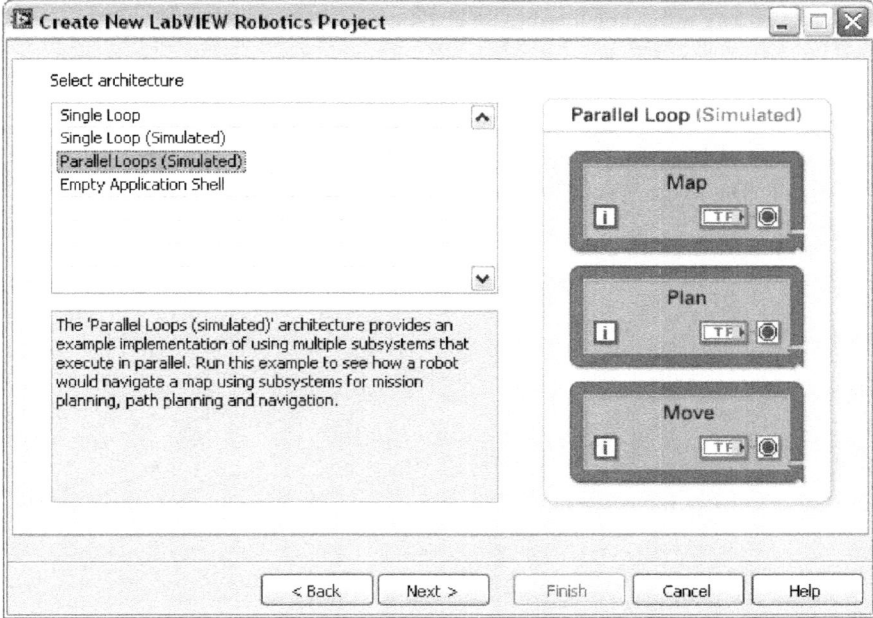

Fig. 2. Sample architectures provide guidance for single and parallel loop applications

The Robotics Project Wizard also provides a more advanced nested control loops architecture that uses multiple timed loops to handle different aspects of robot control. For instance, the timed loops in this example perform the following tasks:

- Mission planning – Returns a random goal position that the robot navigates
- Path planning – Searches a map of the robot environment to find a path to the goal position
- Driving – Simulates the robot moving along the path to reach the goal position

An advanced robot architecture may also include timed loops that detect obstacles, control movement, and measure the progress of the robot. Loops can run on different hardware targets, so you must implement communication across the application. For instance, the nested control loops example uses shared variables to communicate data between timed loops in different VIs.

3   Conclusion

LabVIEW Robotics provides all of the necessary software tools needed for interfacing with sensors and actuators, controlling robot's motion, implementing navigation algorithms and advanced control, and simulating robots in a dynamic environment. A LabVIEW Robotics Starter Kit is available that provides a low-cost entry into LabVIEW Robotics and RIO hardware as well as a fully assembled four-wheeled robot complete with an ultrasonic sensor, TETRIX motors, and two encoders.

References

1. Gretlein, S.: Introducing LabVIEW Robotics – From Fantasy to Fact, INL Q1 (2010)
2. Kopp, E.: The Robot Revolution - LabVIEW Addresses the Needs of an Emerging Market, INL Q4 (2009)
3. Jamal, R., Heinze, R.: Virtuelle Instrumente in der Praxis 2010. VDE Verlag (2010)

Estimation of Mobile Robot Pose from Optical Mouses

Lenka Mudrová, Jan Faigl, Jaroslav Halgašík, and Tomáš Krajník

The Gerstner Laboratory for Intelligent Decision Making and Control,
Department of Cybernetics, Faculty of Electrical Engineering
Czech Technical University in Prague
{mudrol1,halgajar}@fel.cvut.cz,
{xfaigl,tkrajnik}@labe.felk.cvut.cz

Abstract. This paper describes a simple method of dead-reckoning based on off-the-shelf components: optical mouses and a laptop. The problem is formulated as finding a transformation of mouses positions to position of the robot. The formulation of the transformation is based on a method already used in range-based localization. Beside a solution of the transformation, the paper provides description of practical application of mouse based localization for a home made robot. The paper considers identification and mouse data reading procedures as well. The presented approach has been evaluated in several real experiments and the proposed localization provides competitive results to the odometry based on high-precision stepper motors.

1 Introduction

An optical computer mouse used to control a cursor on the screen can inspire to create an open-loop system, which estimates position of a mobile robot. The mouse provides local changes of its sensor position in two directions, so it is clear that estimation of robot pose (position and orientation) cannot be based just on one mouse. Either two or more mouses have to be used or additional information (robot kinematic constrains, compass data) must be taken into account. An application of the mouse to the self-localization problem of a mobile robot came out almost immediately with introduction of an optical mouse. Since then the problem has been studied for several years and successful applications of mouses-based robot self-localization have been reported by many authors. The basic problem formulation for one mouse can be based on a velocity equation [4]. If several mouses are used, detection of wrong measurements can be considered [9]. Precision of localization can also be increased by combination with other sensors [8]. The precision of the optical mouse depends on a surface on which a mouse is moved. Authors of [6] examined several surfaces and reported problems of non-straight displacements. To increase precision of pose estimation, several mouses can be combined and rigid-body constraints can be considered to verify consistence of data measurements [2].

D. Obdržálek and A. Gottscheber (Eds.): EUROBOT 2010, CCIS 156, pp. 93–107, 2011.
© Springer-Verlag Berlin Heidelberg 2011

In this paper we formulate the self-localization as a problem of finding transformation between two robot positions. The transformation is composed from rotation and translation, which is principally same as a method already used in the aforementioned papers. The main difference of the proposed solution resides in computation of the transformation parameters that does not require calculation of inverse (pseudo-inverse) matrix, which can suffer from numerical issues. Our formulation leads to a solution already used in localization methods that are based on range measurements. It allows straightforward combination of several mouses as well as detection of possibly wrong measurements that are discarded. To our best knowledge such approach has not been applied for optical mouses before. An additional contribution of this paper is a presentation of a practical approach, which shows how off-the-shelf components can be combined in a simple and straightforward manner in order to create a cheap independent localization system for a mobile robot. Such a system can be used in EUROBOT competitions, because the playfield is typically flat and uniform, which is suitable for an optical mouse.

The rest of the paper is organized as follows. The next section briefly describes basic principle of mouse sensors and used optical mouses. Problem formulation and pose estimation equation are presented in Section 3. A procedure for identification of sensor parameters is described in Section 4. Practical application of the procedure and realization of mouse based localization system are considered in Section 5, where possible issues are described. Experimental results are presented in Section 6. Described methods and results are discussed and future work is proposed in Section 7.

2   Basic Principle of Optical Mouse Sensor and Used Optical Mouses

The main function of computer mouse is measuring translations in two orthogonal axes. A basic scheme of the optical mouse principle is depicted in Fig. 1. The method is based on correlation of two consecutive images captured by an optical sensor, which is principally low-resolution camera (tens of pixels). Camera resolution together with its optics defines a visible area by the sensor. Two consecutive images have to overlap, otherwise a correlation between two consecutive images cannot be found and the traveled translations are not estimated. This limits maximal measurable speed of a mouse. Precision of a particular mouse is affected by resolution of the sensor and capability to recognize surface details. Surface details recognition can be enhanced by proper lighting, which is usually provided by a light or laser emitting diode. Mouses with laser diodes usually provides higher precision, therefore they are more suitable for mobile robot localization.

Regarding precision, price and size, we decided to use four "Genius NetScroll+ Mini Traveler laser" mouses for our localization system. The mouses use the PAW3601DH-NF optical sensor that provides resolution of 1600 counts per inch and maximal speed of 28 inches per seconds (around 0.7 m.s^{-1}) with acceleration

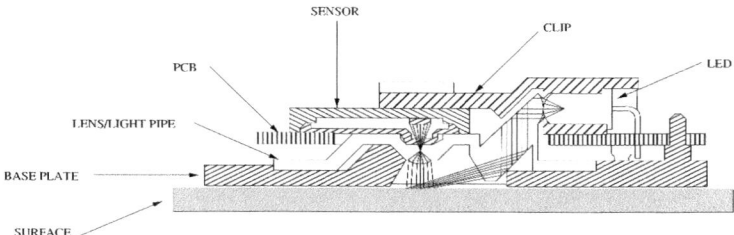

Fig. 1. Mechanism of optical mouse

rates up to 20g. The sensor contains a DSP chip that operates at 27 MHz and processes 6600 frames per second. The mouse can be plugged to USB interface and it does not require a special driver in commonly used operating systems.

3   Estimation of Robot Pose - Dead Reckoning

An optical mouse provides information about position changes in two orthogonal directions defined by the mouse sensor. The changes are denoted as Δx_m and Δy_m. A summation of these changes can be used to estimate position of the mouse. Thus if a mouse is moved from position $(0,0)$ then after several mouse movements (without rotation) the position of the mouse can be estimated as $(\sum_{t=1}^{n} \Delta x_m(t), \sum_{t=1}^{n} \Delta y_m(t))$, where n is the number of position changes sent by the mouse. In order to estimate position of a robot together with its orientation several mouses can be used. Beside the approach described in [8] the basic equation can be formulated in a slightly different way leading to a problem formulation that can be solved by another localization approach called Iterative Closest Point (ICP) [5]. The idea is based on consideration of optical mouse sensor output as a measurement of a position of a point in a plane like range measurements of laser scanners or infrared sensors. The main advantage of mouse sensors is that the correspondence problem is already solved, as data are received from particular mouse.

Assume following basic equation of the robot movement composed of turning about an angle ω and translation about a vector T between time interval $\langle t, t+1 \rangle$:

$$\varphi(t+1) = \varphi(t) + \omega$$
$$X(t+1) = R_{\varphi(t+1)}T + X(t) \tag{1}$$

where φ denotes orientation of the robot, X denotes position of the robot with respect to global coordinate system, $R_{\varphi(t+1)}$ is the rotation matrix with the angle $\varphi(t+1)$. The problem is to find values of ω and T from particular movements of mouses.

3.1   Robot Movement

To describe the idea of the robot pose estimation assume a robot with a local coordinate system O_r. Two mouses are attached to the robot in general position,

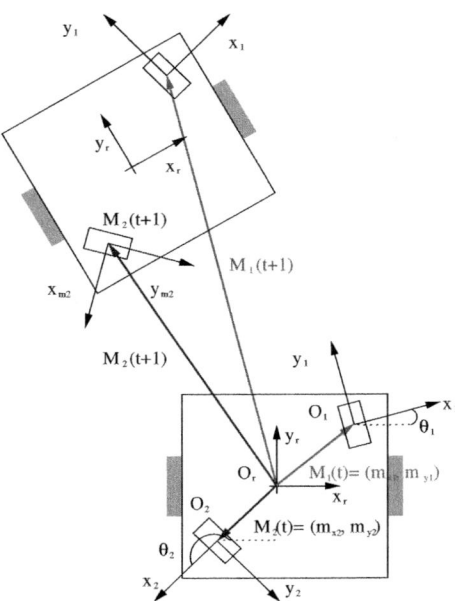

Fig. 2. Robot movement

see Fig. 2. The first and the second mouses define coordinate systems O_1 and O_2 respectively. The centers of the mouses are at positions $M_1(t)$, resp. $M_2(t)$ in O_r, i.e. coordinate system of the robot. The orientation of the mouses with respect to orientation of the robot is θ_1, resp. θ_2. When the robot is moved, mouses centers according to O_r become $M_1(t+1)$ and $M_2(t+1)$. The mouse provides changes of position according to mouse coordinate system, thus position of mouse center i can be estimated by

$$M_i(t+1) = M_i(t) + R_{\theta_i} \begin{bmatrix} \Delta x_i \\ \Delta y_i \end{bmatrix}. \tag{2}$$

A pair of positions $M_i(t)$ and $M_i(t+1)$ is provided for each mouse and the problem is to find a common transformation (i.e. values of ω and T) between points in the pairs. For k mouses a quadratic criterion can be formulated:

$$E(\omega, T) = \sum_{i=1}^{k} |R_\omega M_i(t) + T - M_i(t+1)|^2, \tag{3}$$

which is exactly the same criterion as in the ICP algorithm [5]. A solution of Eq. (3) can be found in an analytical form [5]:

$$\omega = \arctan \frac{S_{xy'} - S_{yx'}}{S_{xx'} + S_{yy'}}$$

$$T = \begin{bmatrix} \overline{x}' \\ \overline{y}' \end{bmatrix} - R_\omega \begin{bmatrix} \overline{x} \\ \overline{y} \end{bmatrix}, \tag{4}$$

where

$$\overline{x} = \frac{1}{k}\sum_{i=1}^{k} x_i(t), \qquad \overline{y} = \frac{1}{k}\sum_{i=1}^{k} y_i(t),$$

$$\overline{x}' = \frac{1}{k}\sum_{i=1}^{k} x_i(t+1), \qquad \overline{y}' = \frac{1}{k}\sum_{i=1}^{k} y_i(t+1),$$

$$S_{xx'} = \sum_{i=1}^{k}(x_i - \overline{x})(x_i(t+1) - \overline{x}'), \quad S_{yy'} = \sum_{i=1}^{k}(y_i - \overline{y})(y_i(t+1) - \overline{y}'),$$

$$S_{xy'} = \sum_{i=1}^{k}(x_i - \overline{y})(y_i(t+1) - \overline{y}'), \quad S_{yx'} = \sum_{i=1}^{k}(y_i - \overline{x})(x_i(t+1) - \overline{x}').$$

$$(5)$$

The estimation of the angle ω and the translation vector T can be performed for each step and new robot position $X(t+1)$ in the global coordinates is then found from Eq. (1).

3.2   Detection of Outliers

If more than two mouses are used, detection of possible wrong measurements can be based on evaluation of error for each particular mouse measurement - Δx_m and Δy_m. The mouse measurement with the highest value of the error, according to Eq. (3), can be discarded and transformation can be recomputed from the remaining values. Alternatively several combinations of measurements from two or more mouses can be computed and a transformation with the lowest overall error $E(\omega, T)$ can be used to estimate the robot pose.

4   Parameters Identification

Each mouse attached to the robot has three basic parameters that have to be identified to use Eq. (4): orientation of the mouse θ_i and position of the center of the mouse sensor to the center of the robot $M_i = (x_i, y_i)$. Besides, a distance conversion parameter from the mouse units to the metric units should be estimated to provide position of the robot in more human intuitive fashion, e.g. in meters or inches. The following identification procedure assumes a robot with differential nonholonomic drive that is capable of straight forward movement by a given distance and rotation by a given angle.

Values of θ_i can be estimated from the forward movement of the robot by a certain distance:

$$\theta_i = \arctan \frac{\sum_{t=1}^{n} \Delta y_i(t)}{\sum_{t=1}^{n} \Delta x_i(t)}, \qquad (6)$$

where $\Delta y_i(t)$, resp. $\Delta x_i(t)$, are particular measurements in time t provided by the ith mouse.

If the traveled distance d (e.g. in meters) is known then the distance conversion parameter can be estimated from the forward movement. Mouses can be placed in general orientation, therefore the traveled distance in the y axis should be estimated from the rotated values:

$$counts\ per\ meter\ for\ the\ i^{th}\ mouse = \frac{1}{d} \sum_{t=1}^{n} (\Delta x_i(t) \cos \theta_i - \Delta y_i(t) \sin \theta_i). \quad (7)$$

Values of M_i can be estimated if a robot with the differential nonholonomic drive is turned about defined angle ω. The orientation of the mouse θ_i is already known, therefore each measurement can be rotated to the robot coordinate system. Kinematic constraints of a differential drive allow following computation, because a robot is assumed to move only in a direction perpendicular to wheel axis [4]. The position of the robot is not changed during rotation, therefore for a single movement about a small angle $\Delta\omega$ positions of mouses with respect to the center of the rotation can be described by:

$$\Delta\omega \begin{bmatrix} -m_{x_i} \\ m_{x_i} \end{bmatrix} = \begin{bmatrix} \cos \theta_i & -\sin \theta_i \\ \sin \theta_i & \cos \theta_i \end{bmatrix} \begin{bmatrix} \Delta x_i \\ \Delta y_i \end{bmatrix}. \quad (8)$$

For a rotation about the angle ω the positions of the mouses can be estimated from the sum of rotated particular changes:

$$\begin{aligned} m_{x_i} &= \frac{1}{\omega} \sum_{t=1}^{n} (\sin \theta_i \Delta x_i(t) + \cos \theta_i \Delta y_i(t)), \\ m_{y_i} &= -\frac{1}{\omega} \sum_{t=1}^{n} (\cos \theta_i \Delta x_i(t) - \sin \theta_i \Delta y_i(t)). \end{aligned} \quad (9)$$

Though precise identifications are presented in [6,4], the above described procedure is straightforward and relatively easy to implement. The procedure is used to estimate parameters of the mouses in experiments presented in Section 6.

5   Practical Issues

This section is dedicated to practical issues that can be considered prior to development of mouse based localization system for a mobile robot. First of all, mouses are primarily used to control a cursor on the screen where a human eye-hand system provides a feedback control loop. The precision of mouse-hand-eye system can lead to an imagination that precise optical mouse can be easily used for robot localization. However, mouses-based self-localization is an open loop system incapable to correct wrong measurements and a naïve application of mouses can lead to a disappointment. This can be avoided by proper consideration of particular assumptions.

5.1   Mounting Mouses at the Robot

One of the fundamental constraints of the mouse application is the height of the sensor above the floor. In order to have the sensor at a constant height, a spring mechanism should press the mouse to the floor. The pressure cannot be too high, because it have to allow the mouse to traverse small bulges. A schema of our pressure system is depicted in Fig. 3a. It is made from thin long plate of aluminium alloy. The *part 1* is made from stronger plate that is bended. It forms a supporting construction for the *part 2* that is the main spring part. The mouses are fixed by a rotating hinge with one degree of freedom. One spring holds two mouses, therefore two springs are mounted to the bottom of the robot, see Fig. 3b. Because the plastic body of the original mouse is used to hold the mouse electronic board, the sensor is at the correct height. The board has to be precisely placed in the plastic body, therefore a glue is applied to fix the board with the body.

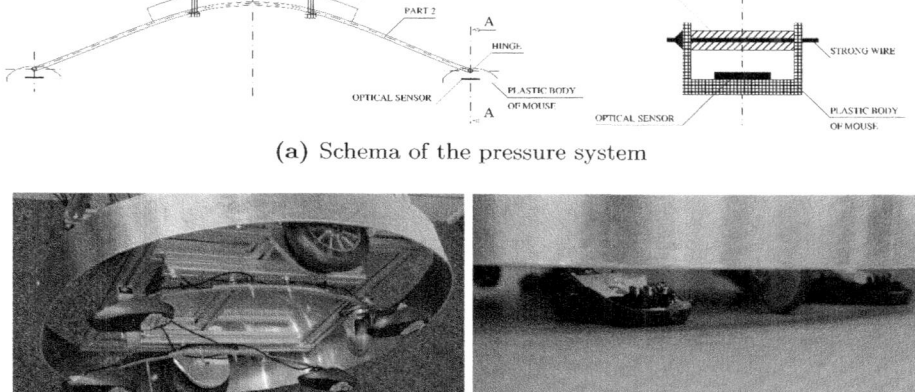

(a) Schema of the pressure system

(b) Attached mouses to the robot chassis

Fig. 3. Mounting mechanism for the mouses

Although usage of four mouses have been planned for the experiments, one of the mouses was not used because of following issue. Accidentally an electronic board of the mouse was twisted and the mouse provided noisy and inaccurate data even if it was placed on the perfect surface, because of wrong height of its sensor.

5.2   Data Acquisition

An additional issue comes out when one tries to read raw data from the mouse. A standard operating system abstracts hardware devices. In a unix based system the raw access to the hardware device is possible by reading from the particular device file in the /dev directory. In operating systems based on the Linux kernel these devices are located in /dev/input/mouseX, where X denotes particular connected mouse. The USB mouses are processed by the *mousedev* driver, which provides emulation of the PS/2 mouse protocol[1]. The protocol uses a sequence of 3 bytes in which particular changes in the x and y directions (i.e. Δx_m and Δy_m) are stored in one signed byte. That means values from -127 to 127 are returned. These values are provided only if the mouse is moved.

If a laptop (or a netbook) is used as the robot on-board computer, the robot movement will cause movement of the cursor in a graphical computer desktop environment, because mouses are typically used to control the cursor. The newly added mouse to the system is automagically attached to the running graphical desktop environment. If the hal daemon is used as a device manager, undesirable cursor movement can be avoided by a policy rule to remove the mouse X-driver, see Listings 1.1. Such a rule can be stored in the file and placed into /etc/hal/fdi/policy directory.

```xml
<?xml version="1.0" encoding="ISO-8859-1"?>
<deviceinfo version="0.2">
   <device>
       <match key="info.product" contains="Genius_Laser_Mouse">
           <remove key="input.x11_driver"></remove>
       </match>
   </device>
</deviceinfo>
```

Listing 1.1. An example of policy rule to disable mouse movement of the cursor

Despite the fact that all mouses should provide measurements at the same moment, so the mouses can be read sequentially, the poll function, which examines a set of file descriptors for an activity, can be used to notify the program about new ready measurement. High speed of nowadays computers causes the poll function is return and data reading is performed only for one mouse at a time. The method described in Section 3 is not suitable for computing robot position change from a single mouse measurement. Therefore data from all mouses are collected before performing position estimation. Alternatively we can wait a certain time period, while particular changes of the mouse positions are accumulated. An example of the reading program is depicted in Listings 1.2, where the real timer is used to notify the program by the signal SIGALRM that data should be processed.

---

[1] The driver can be switch into ImPS/2 or EXPS/2 protocol by sending special magic sequence to the device that can be found in the kernel source in the file mousedev.c.

```
struct  pollfd  ufdr[nbr_mices];
struct  itimerval  timer;
for  (int  i = 0;  i < nbr_mices;  i++) {
    ufdr[i].fd = mouse[i];
    ufdr[i].events = POLLIN | POLLRDNORM;
}
timer.it_interval.tv_sec = period / 1000;  // period is in ms
timer.it_interval.tv_usec = (period % 1000) * 1000;
timer.it_value.tv_sec = period / 1000;
timer.it_value.tv_usec = (period % 1000) * 1000;
signal(SIGALRM, alarm_handler);
setitimer(ITIMER_REAL, &timer, 0);
while (!quit) {
    if (poll(ufdr, nbr_mices, 10)>0) {//poll with 10 ms
        timeout
        for (int i = 0;  i < nbr_mices;  i++) {
            if (ufdr[i].revents & (POLLIN | POLLRDNORM)) {
                read(mouse[i])
    } } }
    if (timer_expires) {
        timer_expires = false;
        process_data(mouse, nbr_mices);
        signal(SIGALRM, alarm_handler);
        setitimer(ITIMER_REAL, &timer, 0);
} } //end read loop
```

Listing 1.2. An example of mouse readings

The period can affect precision of the pose estimation, that is why several experiments have been performed to verify precision of the localization for different values of period.

5.3   Precise Rotation of the Robot

The identification procedure described in Section 4 assumes that the robot is moved in a forward direction by a given distance and it is rotated by the given angle. The traveled distance can be measured manually with sufficient precision, but precise robot rotation is problematic. The following procedure has been performed to rotate the robot about 360°.

1. A laser pointer is attached at the body of the robot.
2. A label of the pointer is placed on a wall that is far enough from the robot.
3. The robot is turned approximately about an angle less than 360°.
4. Then the robot is manually turned to match the laser pointer with the previously placed label on the wall.

The procedure provides sufficient precision for identification of mouses positions M_i. A scene of the robot, label, and laser pointer is shown in Fig. 4.

Fig. 4. Mounted laser pointer at the robot and label on the wall during precise rotation about 360°

6   Experiments

Three optical mouses are used in the self-localization system, because the fourth mouse with the twisted electronic board provided too noisy data. All experiments have been performed with the experimental robotic platform called G^2BOT developed at the Gerstner Laboratory. The platform is based on the ER1 kit from the Evolution Robotics [1] with two stepper motors and the RCM control unit. The robot movements are controlled by the Player program [7]. It should be noted that the RCM is connected to the on-board computer by the USB and together with the Player software leads to a significant transport delay, which causes robot movement to be imprecise. For example, the robot does not move exactly forward when commanded to do so, because one stepper is started a bit earlier than the second one. The imperfections of the robot control loop causes real paths to deviate from desired ones, however a path calculated from RCM board odometry looks like the desired one. This property is not an issue in the performed experiments, because the optical mouses are independent system and the true (real) final positions of the robot have been measured manually.

The proposed method of pose estimation is verified in a set of experiments where the robot is navigated along the path with square shape that is approximately two meters long. The robot moves in turn-move manner, which means the robot is turned in the desired directions at first. After that, a forward motion by the given distance is performed. In our experiments, robot forward velocity was around 0.05 m.s^{-1}, like in [3], and the radial velocity was around 0.05 rad.s^{-1}.

The error of the proposed localization method is evaluated by the distance of the real position from position provided by the localization system at the final point of the path. The average value of the distance from several measurements is denoted as $\bar{d}$. The repeatability of the localization precision is measured as the sample standard deviation of the distances that is denoted as s_n.

The pose estimation is also examined for several reading periods in which particular changes are summed before the position is estimated. In all other cases the estimation is calculated from the data record that contains changes of all used mouses. The robot is primarily moved on a flat surface that is located in the main corridor of the university building. The surface represents typical indoor floor and does not contain any significant disturbances. The identified parameters are then used for the playfield that is made to match the standard surface of the EUROBOT competition, as it is created from the clipboard and the surface is relatively rough.

6.1   Pose Estimation

At first the parameters of the used mouses are identified by the procedure described in Section 4. The robot is moved around 55 cm and turned about 360°. Identified parameters are depicted in Table 1. To simplify transformation from sensor units to cm the average value 662 is used in both sensor axes and for all mouses.

Table 1. Estimated parameters of the optical mouses

nbr. Mouse	θ [°]	m_x [cm]	m_y [cm]	sensor units[2] per cm
1	179.7	7.82	-12.94	642
2	179.6	-8.03	-12.98	663
3	179.8	-8.46	19.10	682

Two selected paths obtained from the mouse localization system are shown in Fig. 5a, where the true positions of the robot are visualized as small '+'.

The precision of the localization from the mouses and inbuilt odometry system of the RCM is depicted in Table 2, where 10 ms and 100 ms are reading periods. It should be noted that estimated path depends on the initial position of the robot, which has been set manually and only approximately to the same place. The deviation is more important than the average value, because it represents repeatability of the localization.

The reading period decreases the precision of the localization. Notice that s_n is similar for both examined periods. It indicates possible same results for higher velocity of the robot, e.g. if the robot will be moved ten times faster. The worse precision can be caused by the naïve summation of measurements without rotation about the angle θ_i. From this perspective estimated position of the robot is not so bad, which can be affected by the turn-move navigation together with the almost parallel orientation of the mouses with the forward direction of the robot. An example of estimated paths for both periods are shown in Fig. 5b.

---

[2] In y direction.

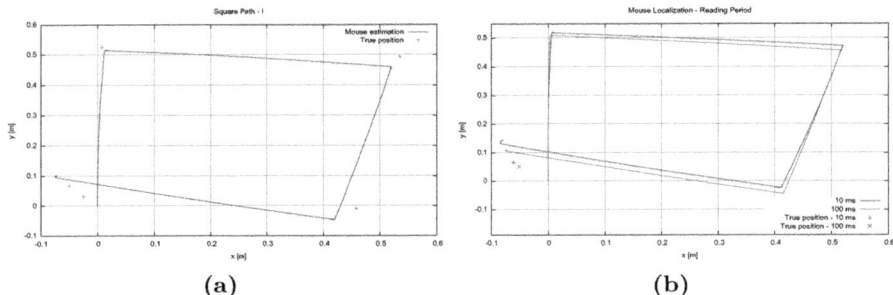

Fig. 5. Selected paths with true positions

Table 2. Precision of the localization method

Method	$\bar{d}$ [cm]	s_n [cm]
mouses, *immediate processing*	2.09	0.77
RCM odometry	5.02	0.68
mouses, period 10 ms	4.02	2.23
mouses, period 100 ms	2.54	2.21

6.2 Playfield Surface

It is a known fact that resolutions of the mouses depends on the surface [6]. To evaluate influence of different surfaces, position of the robot is determined with identified parameters from the prior experiments and compared with the new set of parameters for the playfield that are shown in Table 3. The new average value of the transformation constant of the sensor units per cm is 660.

An example of a path obtained from the proposed method with the prior found parameters and a path determined for the new identified parameters are shown in Fig. 6. The overall localization errors and standard deviations are summarized in Table 4. The standard deviations are very similar, therefore the higher value of $\bar{d}$ is cased by the systematic error that is reduced by the new identification of parameters.

Table 3. Estimated parameters of the optical mouses for the playfield surface

nbr. Mouse	θ [°]	m_x [cm]	m_y [cm]	sensor units^2 per cm
1	179.7	8.40	-13.67	645
2	179.3	-8.80	-13.65	647
3	179.7	-9.69	21.07	685

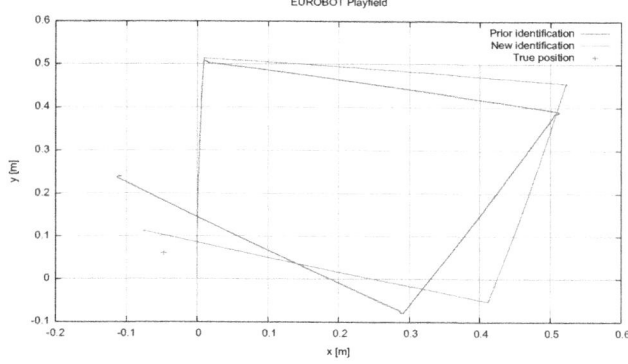

Fig. 6. Differences of the estimated robot position with prior and new identified mouses parameters

Table 4. Overall precision of the localization methods for the EUROBOT playfield

Method	$\overline{d}$ [cm]	s_n [cm]
mouses - *prior identification*	18.69	0.27
mouses - *new identification*	5.45	0.31
RCM odometry	4.88	0.34

6.3   An Example of Outlier Detection

The advantage of using several mouses is the possibility to detect a wrong measurement, i.e. detection of outliers. The effect of outlier detection is demonstrated in Fig. 7. One of the three used mouses is incorrectly identified and

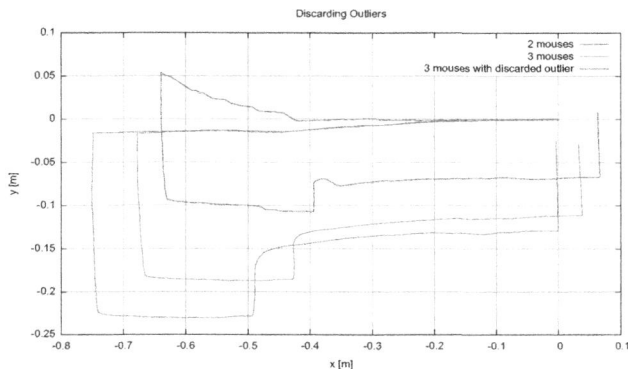

Fig. 7. Estimated robot position from three optical mouses with one incorrectly identified mouse

provides wrong measurements. The red path represents position of the robot computed from one incorrectly and one correctly identified mouse. If all three mouses are used, the position estimation precision is improved, but it is still affected by wrong measurements. The estimation is improved significantly if the outlier detection is used and outliers are excluded from the position calculation.

7   Conclusion and Future Work

The presented results show that used optical mouses provide competitive estimation of the robot position to the precise odometry provided by the RCM. Regarding to the cost of the RCM, the mouses are cheaper, therefore the optical mouses are considered to be useful for a localization of a robot on flat surfaces like in the EUROBOT 2010 competitions. The main advantage of mouse based localization system is its independence on other parts of the robot system, so it can be used in cases where a slippage or a collision affects robot position.

Despite the fact that feasibility of the proposed method and problem formulation has been verified in a set of preliminary experiments, additional experiments should be performed. In these experiments, the following aspects should be considered. A more general path has to be examined instead of the turn-move navigation, e.g. a path composed circular segments. Additional experiments should be performed to estimate the effect of robot velocity on the precision. The pose estimation might be improved if resolution of each mouse is considered individually instead of single conversion parameter, which can be also useful if various types of sensors are combined. The experiments with the reading period should be reconsidered for a more precise model of robot movement, however it seems that immediate data processing provides more accurate estimation of the robot position.

Even though the optical mouses can be easily plugged to an ordinary laptop, an embedded solution based on the set of sensors and microcontroller can be advantageous. Such a system can be smaller and it can require only single (serial) connection to the on-board computer.

Acknowledgments. The support of the Ministry of Education of the Czech Republic, under the Project No. 2C06005, to Jan Faigl is gratefully acknowledged. The support of EU and the Ministry of Education of the Czech Republic, under project No. 7E08006 and ICT-2007-1-216240 to Tomáš Krajník is also acknowledged.

References

1. Evolution Robotics - ER1 Personal Robot System,
   http://www.evolution.com/er1
2. Hu, J.S., Chang, Y.J., Hsu, Y.L.: Calibration and Data Integration of Multiple Optical Flow Sensors for Mobile Robot Localization. In: IEEE International Conference on Sensor Networks, Ubiquitous and Trustworthy Computing, SUTC 2008, pp. 464–469 (2008)

3. Lee, S.: Mobile robot localization using optical mice. In: IEEE Conference on Robotics, Automation and Mechatronics, vol. 2, pp. 1192–1197 (December 2004)
4. Lee, S., Song, J.B.: Mobile robot localization using optical flow sensors. International Journal of Control, Automation, and Systems 2(4), 485–493 (2004)
5. Lu, F., Milios, E.: Robot pose estimation in unknown environments by matching 2d range scans. Journal of Intelligent and Robotic Systems 18, 249–275 (1994)
6. Palacin, J., Valgaon, I., Pernia, R.: The optical mouse for indoor mobile robot odometry measurement. Sensors and Actuators A: Physical 126(1), 141–147 (2006)
7. The Player Project - Free Software tools for robot and sensor applications, http://playerstage.sourceforge.net/
8. Sekimori, D., Miyazaki, F.: Self-localization for indoor mobile robots based on optical mouse sensor values and simple global camera information. In: IEEE International Conference on Robotics and Biomimetics (ROBIO 2005), pp. 605–610 (2005)
9. Sekimori, D., Miyazaki, F.: Precise dead-reckoning for mobile robots using multiple optical mouse sensors. In: Informatics in Control, Automation and Robotics II, pp. 145–151 (2007)

Performance Comparison of Vision Sensors and Processing Power of Two Robotic Platforms for Obstacle Avoidance

Sanda Paturca, Dan Novischi, and Constatin Ilas

University Politehnica of Bucharest,
Splaiul Independentei 313,
Bucharest, Romania
Dan.Novischi@gmail.com

Abstract. This paper presents a comparison of performance for the vision sensors and processing power of two, widely used, robotic systems: NXT Mindstorms and SRV-1 Blackfin. The performance analysis was done in relation to an obstacle avoidance algorithm implemented on both platforms. Three case studies were performed: in the first case study we analyzed the accuracy of object recognitions, in the second case we analyzed the performance of multiple object detection and in third case determined the average time it took each robot avoid the obstacle. These case studies show how much the performance depreciates with the use of low level hardware, like the NXT, for the task of obstacle avoidance.

Keywords: obstacle avoidance, NXT, SRV-1 Blackfin, camera, blob.

1 Introduction

Mobile autonomous robots are becoming increasingly important in people lives. Today, these robots are used in various applications like household cleaning, self driving vehicles, intelligent security systems and space exploration, for the expected benefits of their improved efficiency. In these applications robots have to deal with unstructured environments that contain lots of uncertainties, that is, environments for which there is no prior knowledge of the landscape and the locations or shapes of the obstacles. Obstacle avoidance is of central importance to these autonomous robot applications. The ability to avoid obstacles gives robots the necessary intelligence to make optimal decisions while traveling in such environments.

Today there are numerous robotic platforms that are used for educational and research proposes, such as: VEX Robotic Kit, Bionoid Robotic Kit, Arduino, NXT Mindstorms Kit, POB BOT, SRV-1 Blackfin Robot, etc...

This paper focuses on the performance analysis, in relation to obstacle avoidance, of two robotic platforms, namely: the NXT Minstorms platform from Lego and the SRV-1 Blackfin Robot from Surveyor. We specifically used these two platforms because they are often the first choice in the development of many educational and research projects [20][21][22].

D. Obdržálek and A. Gottscheber (Eds.): EUROBOT 2010, CCIS 156, pp. 108–117, 2011.

We analyzed two main characteristics the performance of the vision sensors (camera) and the processing power in three case studies. These case studies show that a low level hardware robotic system, like the NXT can achieve a relative good performance in relation to obstacle avoidance algorithms based on vision.

Section 2 presents the previous work related to obstacle avoidance. Section 3 describes the hardware of the two robotic systems and the software designed for obstacle avoidance. Section 4 presents the three case studies and section 5 draws the conclusions.

2   Previous Work

Early work on obstacle avoiding algorithms both for ultrasonic sensor and camera based robotic systems focused on different strategies to determine the size and location of the obstacles. This includes the research of Borenstein, Koren [1] and Khatib [2].

Till today one of the most important work was done as a part of EUREKA Prometheus Project [12] by Ernst Dickmanns. This project was the largest R&D European project ever in the field of self driving cars at very high speeds. The project achievement was the basis for most subsequent work on self driving cars.

Today, academic and industry research focuses on obtaining reliable and accurate solutions, running on simple hardware. To stimulate these activities, a large variety of competitions and platforms address the problem of obstacle avoidance for different purposes. Most significant competitions and platforms are:

- DARPA Grand Challenge [13] – is a competition between self driving cars sponsored by the Defence Advanced Research Projects Agency (DARPA) and provides some good, and bad, examples of the current state of the art in obstacle avoidance.

iRobot [16] – is a robotic platform developed by the company with the same name that is used to implement a series of household cleaning, indoor and outdoor, autonomous robots.

3   Hardware and Software

The robot platforms used to perform the experiments are: the NXT Mindstorms Kit with an additional NXT Cam v.2 and the SRV-1 Blackfin Robot. Both robots have a wide range of sensors and motors as actuators.

3.1   Hardware

The NXT Mindstorms Kit [7] is composed of the following parts: a microcontroller, named NXT Brick, three DC motors and four sensors.

The NXT Brick has a 32-bit ARM7 CPU running at 48 MHz, an 8-bit AVR Co-Processor running at 4 MHz that interfaces the I/O ports via I2C protocol with the ARM7 CPU, 4 input ports and 3 output ports. It also provides USB and Bluetooth support to communicate with a PC or a third party device.

The three DC motors are 9V electric motors that can achieve a maximum rotation speed of 200 rpm. Each motor has a built-in rotation encoder. The rotation encoder measures motor rotations in degrees or full rotations with an accuracy of +/-1 degree.

The NXT Camera from Mindsensors [15] is a real-time image processing subsystem. It can track an image of 88 x 144 pixels at 30 frames/second. The subsystem is a self contained unit capable of tracking up to 8 different objects of a pre-specified color. The main advantage of the vision subsystem is that it requires minimal processing resources from the NXT Brick, thus minimizing robot processing requirements.

Our robot, shown in Figure 1, was built using NXT Kit and incorporates the NXT Brick, the NXT Camera and two DC motors mounted differentially.

Fig. 1. The NXT Robot

The SRV-1 Blackfin Robot [6] is a complete hardware system for our proposes. The robot is composed of: a microcontroller, four DC motors, a camera sensor, a wireless communication module and various sensors and expansion boards.

The microcontroller has an blackfin DSP BF537 CPU running at 500MHz with the capability of processing 1000 integer MIPS and 32 MB SDRAM.

The four DC motors are mounted differentially and are controlled via a motor control module which has the capacity of 1A per motor.

The camera built in the robot is an Omnivision OV9655 1.3 megapixel sensor. It can track an image of 160x128 up to 1280x1024 pixels at an average rate of 10 frames / second. This system is not self contained and it relies on the Blackfin CPU for processing the image.

The wireless communication module provides support for communication with a PC or a third party device over TCP/IP or UPD protocols.

The robot is shown in the figure below:

Fig. 2. The SRV-1 Blackfin robot

3.2 Software

To evaluate the two robots performance we designed and implemented a simple obstacle avoidance algorithm based on color recognition. For the NXT robot this algorithm was implemented in Java programming language using leJOS 0.7 firmware [17] and for the SRV-1Blackfin robot [6] the algorithm was implemented in C language by recompiling our modified version of the standard firmware. A known fact concerning the programing languages used to implement our obstacle avoiding algorithm is that, in general, the code written in Java is by a constant factor slower then the code written in C. However the JVM specifically designed for the NXT is a reduced JVM that runs as fast as the code written C based programing languages for this platform [17][19].

The obstacle avoidance algorithm [10][11] was structured in three individual tasks, namely: the detection task, the processing task and the travel task. The main program logic, that reunites these tasks, starts from the idea of robot movement on three segments: from the start point to an obstacle, from the obstacle to a determined intermediate point and from the intermediate point to the final point.

The detection task's goal is to identify an obstacle on the robot travel path based on color recognition in form of blobs. A blob is here a rectangular shape with corresponding color to the obstacle in question. We obtained the data for each of blobs by using the built-in RGB (R – red, G – green, B - Blue) filter functions for both of the robots camera sensors. This data, returned by both built-in functions, is the number of blobs, the top left (x,y) and the bottom right (x,y) coordinates of the blobs. Further data processing, for correct obstacle identification, implemented on both of the robots, was done according to the algorithm presented below:

```
detection Task
  1.while(true)
      1.1. get number of blobs
      1.2. if(number of blobs > 0)
              1.2.1. get the blobs coordinates;
              1.2.2. compute the blobs areas;
              1.2.3. determine the biggest blob;
              1.2.4. determine blobs with 90% of the area
                     overlapping the biggest blob;
              1.2.5. merge biggest blob with the 90% area
                     overlapping blobs;
              1.2.6. set obstacle coordinates with the
                     coordinates of the merged blobs;

              1.2.7. if(obstacle area > threshold)
                 1.2.7.1. notify the travel task;
                 1.2.7.2. while(!finished processing)
                     1.2.7.2.1. wait();
```

The travel task drives the robot between two points by controlling the actuators (motors) and the processing task is responsible for the decisions made by the robot to avoid the obstacle. This processing task determines the relative position of the obstacle to the robot and then it computes the shortest path to avoid it. The algorithm for the processing task is presented below:

```
processing Task
    1. get x left obstacle coordinate;
    2. get x right obstacle coordinate;
    3. compute obstacle width;
    4. compute distance to the obstacle;
    5. compute obstacle relative position to the robot;
    6. compute the shortest path to avoid the obstacle;

    7. finished processing = true;
    8. notify the detection task;
```

The three tasks are synchronized according to the diagram below:

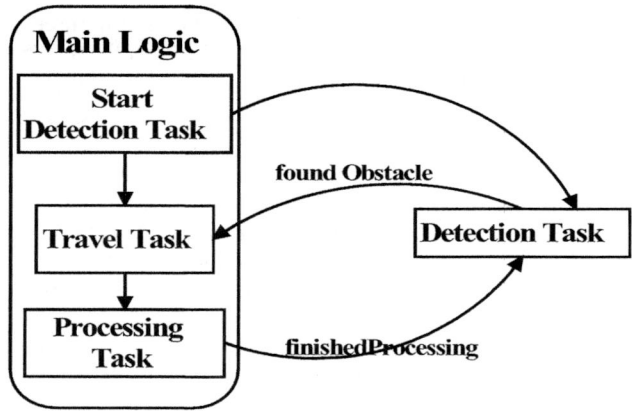

Fig. 3. Synchronization diagram

Although the hardware of the two robots is quite different, for the obstacle avoiding algorithm, we matched the flowing characteristics: the two robots motor speeds and the the distance from which the detection task signals the travel task that an obstacle was found.

4 Performance Comparison

The performance of each robot, from the obstacle avoidance point of view, depends on various factors. The most important factors are: processing power, the camera sensor, the environment surrounding the robot and the object recognition algorithm used. These factors are strongly dependent on each other.

In our first case study we tested the behavior of the two cameras relative to the environment and object recognition algorithm. To accomplish this task we positioned both of the robot cameras at the same distance from a red object (a red coke can) as the object to be recognized and analyzed the vision sensors behavior in two environmental lighting conditions:

- one in which the light intensity was strong and uniform (very little variations given by other objects shadows)

- the other in which light intensity was low, where the environment was illuminated by an incandescent light bulb of 75W

For the two illumination conditions we captured one hundred frames containing the processed images (blob images) from which computed the coordinates of blob's center of gravity (COG) and its area. We computed the variance of the x and y coordinates of the COG and the blobs area. The results are presented in the table below:

Table 1. The variance of the x, y COG coordinates and blob area variance for the two robot cameras in two lighting conditions

Camera	Light Level	Variance of x coordinate	Variance of y coordinate	Variance of blob area
NXT		13.798	24.044	2591.4
Blackfin	High	1.549	1.549	841.47
NXT		24.7667	52.5962	6478.5
Blackfin	Low	3.436	3.237	1121.38

This test shows that the variance of the parameters for the NXT camera is much bigger that the variance of the parameters of the Blackfin camera. We conclude that the NXT camera is sensitive the lightning condition. Due to this sensitivity, object

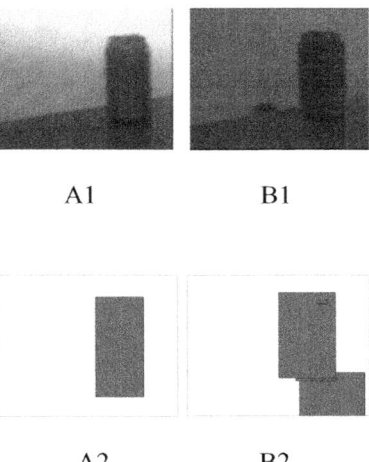

A1   B1

A2   B2

Fig. 4. NXT camera real and processed images for obstacle identification: A1 (real image) and A2 (processed image) – in strong light conditions; B1 (real image) and B2 (processed image) – in low light conditions

recognition was performed with a low accuracy in low light conditions. The Backfin robot camera showed a higher performance than the NXT camera and was able to recognize the object more accurately. Camera real images and blob images for the two cameras in both lighting conditions are presented in figure 5 and figure 6.

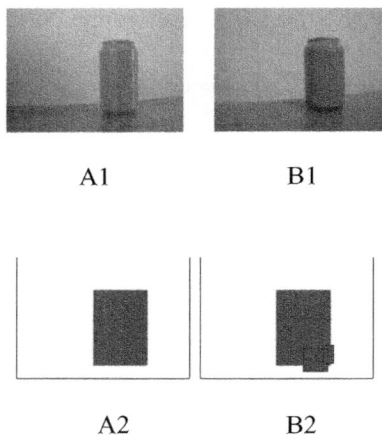

A1   B1

A2   B2

Fig. 5. SRV-1 camera real and processed images for obstacle identification: A1 (real image) and A2 (processed image) – in strong light conditions; B1 (real image) and B2 (processed image) – in low light conditions

In the second case study we tested to see how each of the two cameras together with the object recognition algorithm handled multiple objects detection. For this propose we used two objects of the same color (red coke cans) and we positioned the two robot cameras at the same distance from the objects. We measured the percentage of the cases when the blobs for the two objects overlap on x coordinate. From table 3 we can see than the percentage of the overlapped objects is significantly higher than the Blackfin. As shown in figure 7, when the NXT robot encounters two objects of the same color that are close together, the NXT camera cannot distinguish between the two. In the same case the Blackfin robot camera successfully identified the two objects.

Table 2. Percentage of overlapping x coordinate for the two robotic platforms

	NXT Camera	SRV-1 Blackfin Camera
Overlapping x coordinate percentage	84 %	6 %

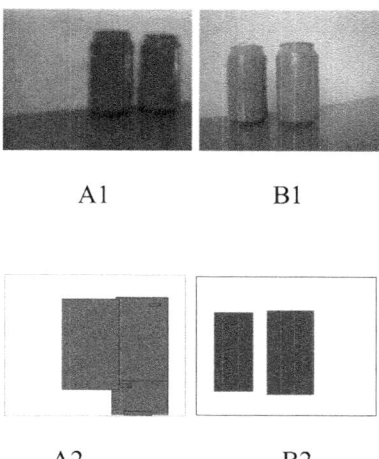

A1   B1

A2   B2

Fig. 6. Robot processed images, where the robots encountered two objects of the same color: A1 (real image) and A2 (processed image) – NXT robot; B1 (real image) and B2 (processed image) – SRV-1 Blackfin robot

In the third case study we tested how much time it took each robot to identify the obstacle (a red coke can) and make a decision to avoid the obstacle. This aspect is dependent on the processing power, the detection task and processing task, described in the previous section. The graphic in figure 8 shows time per distance traveled by the robots. From this figure we can see that the Blackfin robot is much faster in decision making then the NXT robot.

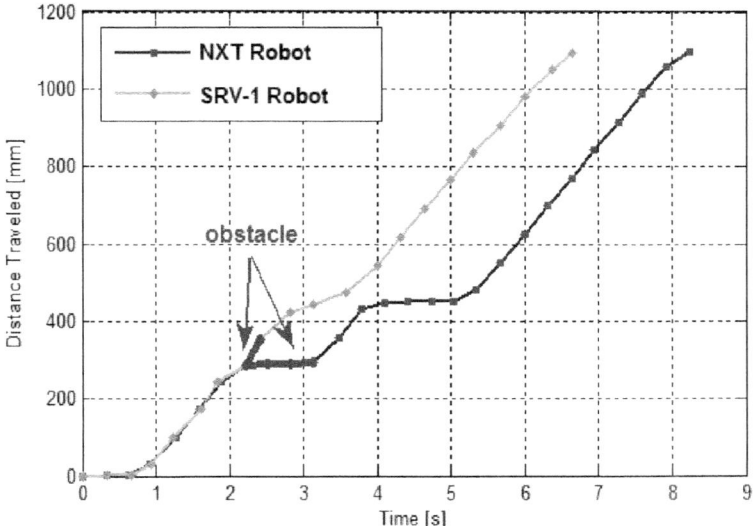

Fig. 7. Traveled distance versus time

Table 3 presents the performance the processing power in MIPS , camera image size and average processing time of the two robotic systems. The results show that the processing time for obstacle avoidance of the NXT robot is approximately three times longer than the SRV-1 Blackfin robot.

Table 3. Processing power, analyzed image size and average processing time to avoid the obstacle for the two robotic platforms

Robot	Processing Power [MIPS]	Camera Image Size [pixels]	Average Processing Time to Avoid the Obstacle [s]
NXT	2	88x144	1.3
SRV-1 Blackfin	1000	160x128	0.4

5   Conclusions

This paper presents a comparison of performance for the vision sensors and processing power of two robotic systems, namely NXT Mindstorms and SRV-1 Blackfin. The performance analysis was done in three case studies in the light of an obstacle avoidance algorithm for low level hardware systems, which we designed and successfully implemented on the two robotic platforms.

The vision sensor for NXT robot that we used was an NXT camera from Mindsensors. This camera is a self contained vision subsystem for real-time image processing. For the SRV-1 Blackfin robot we used the built in Omnivision OV9655 camera.

The algorithm for obstacle avoidance was structured in three individual tasks, namely: the detection task, the travel task and the processing task. The main idea of this algorithm is travel path segmentation. The object recognition for this algorithm was based on color recognition (blob tracking).

In first case study we computed the accuracy with which the cameras of the two robots recognized objects. The accuracy is based on the variance for the blob's center of gravity x, y coordinates and the variance of the blob area. These measurements showed that the NXT camera is more sensitive than the SRV-1 Blackfin camera.

The second case study presents how the robot cameras handled multiple object detection. The performance of the cameras was determined by computing the percentage of overlapping x blob coordinates. The results show that when two objects of the same color are close together the NXT camera cannot distinguish between the two.

The third case study we analyzed the processing power by computing the average time it took each of the two robots to detect and make a decision to avoid an obstacle. This analysis shows that NXT robot has lower performance than the SRV-1 Blackfin robot.

From the three case studies we conclude that: although SRV-1 Blackfin robot has higher overall performance due to its significantly higher processing power and vision sensor, the low level hardware NXT robot achieves a relative good performance for the task of obstacle avoidance.

Therefore for applications that involve obstacle avoiding algorithms based on simple object recognition, as blob detection, with high environmental constrains, as lighting conditions, and that are not strongly time dependent the low level hardware NXT platform can offer similar performance as a high performance robotic platform like the SRV-1.

References

1. Borenstein, J., Koren, Y.: Obstacle Avoidance with Ultrasonic Sensors. IEEE Journal of Robotics and Automation 4 (April 1988)
2. Khatib, O.: Real-time Obstacle Avoidance for Manipulator and Mobile Robots. International Journal Robotics, Res. 5(1), 90–98 (1986)
3. Michels, J., Sexna, A.: High Speed Obstacle Avoidance using Monocular Vision and Reinforcement Learning. Autonomous Systems Laboratory, CIRO ITC Center
4. Jones, J.: Robot Programming: A Practical Guide to Behavior-Based Robotics. McGraw-Hill, New York (2004)
5. Arkin, R.C.: Behavior-Based Robotics. MIT Press, Cambridge (1998)
6. Surveyor Co., http://www.surveyor.com (viewed on February 2010)
7. NXT Mindstorms Kit, http://www.mindstorms.com (viewed on February 2010)
8. NXT Cam View, http://nxtcamview.sourceforge.net/ (viewed on February 2010)
9. Davies, E.R.: Machine Vision: Theory, Algorithms, Practicalities, 3rd edn. ISBN-13: 978-0-12-206093-9
10. Novischi, D.: Autonomous Mobile Robots Avoiding Obstacles using Ultrasonic Sensors or Video Camera. In: The Spot 2009 Proceedings (2009) ISBN: 978-87-89179-84-1
11. Novischi, D., Ilas, C.: Autonomous Mobile Robot Avoiding Obstacles using Ultrasonic Sensor and Video Camera. Robotica & Management 14(2) (2009)
12. Eureka Prometheus Project, http://en.wikipedia.org/ (viewed on December 2009)
13. DARPA Grand Challenge, http://www.darpa.mil (viewed on December 2009)
14. Dudek, G., Jenkin, M.: Computational Principles of Mobile Robotics. Cambridge University Press, Cambridge (2003)
15. Mindsensors, http://www.mindsensors.com (viewed on February 2010)
16. IRobot, http://store.irobot.com (June 2009)
17. Lejos – Java for Lego Mindstorms, http://lejos.sourceforge.net/ (viewed on February 2010)
18. Ekvall, D., Kragic, D., Jensfelt, P.: Object detection and mapping for service robot tasks. Robotica: International Journal of Information, Education and Research in Robotics and Artificial Intelligence (2007)
19. NXT Programing Software, http://www.teamhassenplug.org/NXT/NXTSoftware.html (viewed on April 2010)
20. Odeh, S., Faqeh, R., Abu Eid, L., Shamasneh, N.: Vision-Based Obstacle Avoidance of Mobile Robot Using Spatial Model. American J. Engineering and Applied Sciences 2(4), 611–619 (2009)
21. Certo, J., Cordeiro, N., Reinaldo, F., Reis, L.P., Lau, N.: A Tool for Evaluating Teams' Performance in RoboCup Rescue Simulation League. In: Gelbukh, A., Reyes-Garcia, C. (eds.) Special Issue: Advances in Articial Intel-ligence, Research in Computing Science, vol. 26, pp. 137–148 (November 2006) ISSN: 1870-4069
22. Simpson, J., Ritson, C.G.: Toward Process Architectures for Behavioural Robotics. In: Welch, P.H., et al. (eds.) Communicating Process Architectures. IOS Press, Amsterdam (2009)

Combining Gaussian Processes and Conventional Path Planning in a Learning from Demonstration Framework

Markus Schneider, Richard Cubek, Tobias Fromm, and Wolfgang Ertel

Autonomous Mobile Service Robots Laboratory
Ravensburg-Weingarten University of Applied Sciences
Doggenriedstrasse, 88250 Weingarten, Germany
firstname.lastname@hs-weingarten.de
http://www.zafh-servicerobotik.de/en

Abstract. Today, robots are already able to solve specific tasks in laboratory environments. Since everyday environments are more complex, the robot skills required to solve everyday tasks cannot be known in advance and thus not be programmed beforehand. Rather, the robot must be able to learn those tasks being instructed by users without any technical background. Hence, Learning from Demonstration (LfD) is one of the essential topics to bring robots out of the lab moving towards everyday robustness. The key property of an agent regarding a demonstration learned skill is its ability of generalization, that is, applying a learned skill to situations that differ from those during demonstration. In this paper, we present a method to enhance the generalization capabilities of an advanced new LfD framework by combining it with conventional path planning.

Keywords: Learning from Demonstration, Programming by Demonstration, Gaussian Processes, Path Planning.

1 Introduction

Preparing robots to face everyday environments requires them to have a set of predefined abstract skills, where abstraction is needed to generalize over similar tasks. But is it possible to predefine all skills a robot needs during lifetime? For example, setting the table consists of several simple *pick and place* tasks, which the robot should be able to handle natively. But additionally, there are underlying constraints such as the necessity of placing saucers on the table before placing the cups, to mention only one.

Taking a closer look at the problem, the conclusion is that it is impossible to predefine all needed skills for complex everyday environments. Robots will have to acquire them during lifetime fulfilling the requirement of reproduction in situations that differ from those during demonstration. Such a skill aquisition could be realized in different ways. Programming or describing by any kind of programming or description language is not suitable for daily use, especially

D. Obdržálek and A. Gottscheber (Eds.): EUROBOT 2010, CCIS 156, pp. 118–129, 2011.

not for users without profound computer knowledge. Verbal robot instruction is an interesting, but comparatively small research area [Bug03] and should be treated as an extension to other techniques. Since even humans tend to learn by following examples rather than verbal instructions, many research groups follow a Learning from Demonstration (LfD) approach. This important field is also addressed by on of the robocup@home tasks, the *CopyCat* game, where the robot has to reproduce a human demonstrated block movement in a game-like setting.

1.1   Learning from Demonstration: Different Approaches

In general, the representation of a skill can take place on two abstraction layers: a low level representation *(trajectory encoding)* for generic motions and a high level representation *(symbolic encoding)* for sequences of predefined actions [BCDS08]. Trajectory encoding allows us to describe arbitrary motor primitives and during training very often requires direct motion of the robot's actuator by a human trainer. High-level learning requires a predefined set of low-level skills, but allows the description of a more abstract action sequence for more complex, goal oriented tasks. LfD can address different abstraction layers. For instance, a high-level learning approach can be based on the detection of spatio-temporal task constraints, as presented in [EK08].

Since motor primitives have to be mastered before high-level skills can be applied, most works are located in the area of low-level skills. A popular approach is a demonstration of a motion by a human, followed by an optimization via Reinforcement Learning. This area of research is mainly focused on advanced robot actuators of high mechanical flexibility for handling low-level tasks which are very difficult to control even for humans. Interesting examples are the *ball in cup* [KMP08] and *t-ball batting* [PS07] problems.

There are also approaches to develop an integrated overall solution as presented in [EZRD02]. A demonstration is observed by a set of distributed sensors, recording motions (including graspings) as well as object positions to allow a detailed analysis of the human behavior. Recognizing elementary actions, the demonstration is divided into semantically related segments and mapped onto a sequence of predefined symbols. After further processing, this abstract representation can be mapped onto low-level skills of suitable robotic systems realizing both, reproduction of complex skills on changing environments and changing robotic embodiments.

Another powerful technique is the low-level time-depending observation of the actuator's end effector relative to each object in the task [Cal07, CB08]. After a couple of demonstrations, the time-depending variations between demonstrations are modelled by Gaussian Mixture Regression (GMR). In the reproduction phase, the generated trajectories relative to each object are weighted anti proportional to their variance. The reproduction can now be generated in compliance to the variations, successfully handling new situations where the objects can be placed at arbitrary positions. The first part of our work is based on this approach with the difference that we use Gaussian Processes (GP) instead of GMR.

1.2   Generalization Issue

Beside robustness, a demonstration learned behavior is evaluated by its applicability to situations that differ from those during demonstration. The abstraction capability of a GP based LfD framework is strong. After three or four demonstrations, the robot will solve the demonstrated task successfully in a scene with objects being repositioned arbitrarily. As a weakness of this method, the generalization capability ends at the point where objects or obstacles occur which have not been present or relevant during demonstration. This kind of objects is ignored by the motion plan what may cause collisions. We address this problem with our extended and combined method. It enhances the learned skill by expanding the class of situations the skill can be applied to.

2   Methods

2.1   Gaussian Process Models

In recent years, Gaussian Processes became very popular in the context of machine learning for regression and classification problems [RW06]. They are powerful tools to solve non-linear problems while requiring relatively simple linear algebra only. The Gaussian Process theory provides a very natural way to define a prior distribution over (regression) functions. In the standard non-linear regression problem, we try to estimate a latent function $f(\boldsymbol{x})$ given input variables $\boldsymbol{x} \in \mathbb{R}^D$ and noisy observed target values $y \in \mathbb{R}$ modeled as $y = f(\boldsymbol{x}) + \epsilon$, where $\epsilon \in \mathbb{R}$ is a random noise variable that is independent, identically distributed (i.i.d.) for each observation. The training data set comprising n input points together with the corresponding observations is denoted by $\mathcal{D} = \{(\boldsymbol{x}_i, y_i)|_{i=1}^n\}$. The Gaussian Process regression model tries to learn the predictive distribution $p(f^*|\boldsymbol{x}^*, \mathcal{D})$ of a new test output f^* given a test input $\boldsymbol{x}^*$. To simplify notation, all training inputs $\{\boldsymbol{x}_i\}_{i=1}^n$ are collected in a so called *design matrix* $\boldsymbol{X}$ of size $D \times n$ and we define the matrix $\boldsymbol{X}^*$ and the vectors $\boldsymbol{f}^*, \boldsymbol{y}$ in the same way. The key in Gaussian Processes is to consider the training outputs $\boldsymbol{y}$ and the new testing points (prediction values) $\boldsymbol{f}^*$ as a sample from the same (zero mean) multivariate Gaussian distribution. The predictive distribution is then again a multivariate Gaussian distribution $\mathcal{N}(\bar{\boldsymbol{f}}^*, \mathrm{cov}[\boldsymbol{f}^*])$ conditioned on the training data with mean

$$\bar{\boldsymbol{f}}^* = \boldsymbol{K}^*(\boldsymbol{K} + \sigma_n^2 \boldsymbol{I})^{-1}\boldsymbol{y} \tag{1}$$

and covariance matrix $\mathrm{cov}[\boldsymbol{f}^*]$

$$\mathrm{cov}[\boldsymbol{f}^*] = \boldsymbol{K}^{**} - \boldsymbol{K}^*(\boldsymbol{K} + \sigma_n^2 \boldsymbol{I})^{-1}\boldsymbol{K}^{*T}, \tag{2}$$

where $\boldsymbol{I}$ is the identity matrix, $\boldsymbol{K} \in \mathbb{R}^{n \times n}$, $\boldsymbol{K}_{i,j} = k(\boldsymbol{x}_i, \boldsymbol{x}_j)$, $\boldsymbol{K}^* \in \mathbb{R}^{n^* \times n}$, $\boldsymbol{K}_{i,j}^* = k(\boldsymbol{x}_i^*, \boldsymbol{x}_j)$ and $\boldsymbol{K}^{**} \in \mathbb{R}^{n^* \times n^*}$, $\boldsymbol{K}_{i,j}^{**} = k(\boldsymbol{x}_i^*, \boldsymbol{x}_j^*)$. The function $k(\cdot, \cdot)$ is called covariance function, or kernel, and is used to construct the covariance matrices $\boldsymbol{K}, \boldsymbol{K}^*, \boldsymbol{K}^{**}$. We will write the Gaussian Process as $\mathcal{GP}(\boldsymbol{0}, k(\cdot, \cdot))$ or simply $\mathcal{GP}$.

Typically, the covariance function depends on parameters $\boldsymbol{\omega}$ which are called *hyper parameter* because they are determined in advance and not used to absorb the training data information. A widely used covariance function is given by the exponential of a quadratic form, namely the *Squared Exponential* (SE) to give

$$k(\boldsymbol{x}_p, \boldsymbol{x}_q) = \sigma_f^2 \exp\left(-\frac{\|\boldsymbol{x}_p - \boldsymbol{x}_q\|^2}{2l^2}\right), \tag{3}$$

with $\boldsymbol{\omega} = (\sigma_n, \sigma_f, l)^T$ defining the *noise level, signal variance* and *characteristic length scale* respectively. We use $p(\boldsymbol{y}|\boldsymbol{X}, \boldsymbol{\omega})$ to obtain the *log marginal likelihood* given by

$$\log p(\boldsymbol{y}|\boldsymbol{X}, \boldsymbol{\omega}) = -\frac{1}{2}\boldsymbol{y}^T \boldsymbol{K}_y^{-1} \boldsymbol{y} - \frac{1}{2}\log|\boldsymbol{K}_y| - \frac{n}{2}\log 2\pi, \tag{4}$$

where $\boldsymbol{K}_y$ substitutes $\boldsymbol{K} + \sigma_n^2 \boldsymbol{I}$ and $|\cdot|$ is the determinant. We then use a *conjugate gradient* algorithm [NW99] to (locally) maximize the log marginal likelihood function with respect to the hyper parameters.

The Gaussian Process Model described so far assumes a constant noise level. Given a data set that requires a variable noise model would be able to correctly estimate the mean value, but it would fail to correctly estimate the variance. We therefore have to introduce a more flexible noise model called *heteroscedasticity*. Replacing the constant noise rate σ_n by an input dependent noise function $r(\boldsymbol{x})$, the mean and covariance function of the predictive distribution changes to

$$\bar{\boldsymbol{f}}^* = \boldsymbol{K}^*(\boldsymbol{K} + \boldsymbol{R})^{-1}\boldsymbol{y} \quad \text{and} \tag{5}$$

$$\mathrm{cov}[\boldsymbol{f}^*] = \boldsymbol{K}^{**} + \boldsymbol{R}^* - \boldsymbol{K}^*(\boldsymbol{K} + \boldsymbol{R})^{-1}\boldsymbol{K}^{*T} \tag{6}$$

respectively. $\boldsymbol{R}$ is defined as $\mathrm{diag}(\boldsymbol{r})$, with $\boldsymbol{r} = (r(\boldsymbol{x}_1), \ldots, r(\boldsymbol{x}_n))^T$, and $\boldsymbol{R}^* = \mathrm{diag}(\boldsymbol{r}^*)$, with $\boldsymbol{r}^* = (r(\boldsymbol{x}_1^*), \ldots, r(\boldsymbol{x}_{n*}^*))^T$. We follow [GWB97] and use a second independent Gaussian Process (the z-process denoted by $\mathcal{GP}_z$) to model the log of the noise level giving $z(\boldsymbol{x}) = \log(r(\boldsymbol{x})) = (z_1, \ldots, z_n)^T$. The z-process maintains its own independent covariance function $k_z(\cdot, \cdot)$ and parameters ω_z. With $\boldsymbol{z}^*$ as the $\mathcal{GP}_z$ posterior prediction, we obtain the predictive distribution $p(\boldsymbol{f}^*|\boldsymbol{X}^*, \mathcal{D}) =$

$$\int \int p(\boldsymbol{f}^*|\boldsymbol{X}^*, \mathcal{D}, \boldsymbol{z}, \boldsymbol{z}^*) p(\boldsymbol{z}, \boldsymbol{z}^*|\boldsymbol{X}^*, \mathcal{D}) \, d\boldsymbol{z} \, d\boldsymbol{z}^*, \tag{7}$$

where the last term prevents an analytical solution of the integral. A common method is to approximate $p(\boldsymbol{f}^*|\boldsymbol{X}^*, \mathcal{D}) \approx p(\boldsymbol{f}^*|\boldsymbol{X}^*, \mathcal{D}, \tilde{\boldsymbol{z}}, \tilde{\boldsymbol{z}}^*)$, where $\tilde{\boldsymbol{z}}, \tilde{\boldsymbol{z}}_* = \arg\max_{\boldsymbol{z}, \boldsymbol{z}^*} p(\boldsymbol{z}, \boldsymbol{z}^*|\boldsymbol{X}^*, \mathcal{D})$. We use the expectation-maximization (EM) algorithm, introduced by [KPPB07], to iteratively estimate $\mathcal{GP}_z$ and then combine it with a noise free $\mathcal{GP}$ to estimate a heteroscedastic Gaussian Process. The whole procedure follows the EM algorithm for the heteroscedastic Gaussian Process Model as described in [KPPB07].

2.2   The GP Based Learning from Demonstration Framework

We use the same learning from demonstration framework as described in [CB08] to encode a task and summarize the used methods. During a demonstration, sensor information is collected together with a timestamp and stored as a state/observation $\boldsymbol{\xi} = (\xi_t^{(i)}, \boldsymbol{\xi}_s^{(i)}) \in \mathbb{R}^D$. Here, $\xi_t^{(i)}$ denoting the i-th *temporal value* $\in \mathbb{R}$ and $\boldsymbol{\xi}_s^{(i)} \in \mathbb{R}^{(D-1)}$ is the i-th vector of *spatial values*. A mapping from $\xi_t^{(i)}$ to $\boldsymbol{\xi}_s^{(i)}$ is called a *policy* and denoted with π. In our experiments we use a simple timestamp t as the temporal value and the spatial value is represented through coordinates in *joint space* $(\theta_1, \ldots, \theta_k$ for the k motor encoders of the robot) or *task space* (Cartesian coordinates and orientation of the end effector). Suppose we have n demonstrations and each demonstration is resampled to a fixed size of T, then the data set $\mathcal{D}$ of all observations will be of length $N = nT$, formally $\mathcal{D} = \{(\xi_t^{(i)}, \boldsymbol{\xi}_s^{(i)}) | i = 1, \ldots, N\}$. The policy used here maps from ξ_t to $\boldsymbol{\xi}_s$ using $p(\boldsymbol{\xi}_s | \xi_t, \mathcal{D}) = \mathcal{GP}(\cdot, \cdot) \forall \xi_t \in \mathbb{R}$ in terms of a Gaussian Process. This approach perfectly fits the requirements of an approximation in the context of robots, namely generating *continuous and smooth paths* and provide a *generalization over multiple demonstrations*. The Gaussian Process covariance function controls the function shape and ensures a smooth path and the fundamental GP algebra calculates the mean over demonstrations. This reduces noise (introduced by sensors and human jitter during the demonstration) and recovers the underlying trajectory.

The key idea is to use the variations and similarities between demonstrations to extract *what* is important to imitate. Assume a robot arm scenario where the goal is to reach a specific end state (e.g. given 3D coordinates) and the start state is chosen randomly. The trajectories will differ significantly at the beginning, whereas the position should be more or less the same at the end. Such a part of a trajectory with low variation is defined as a *constraint* because no discrepancy is desired. The *heteroscedastic* Gaussian Process Model, as discussed in section 2.1, allows us to encode both the trajectory and the variation between the demonstrations.

In the training (encoding) phase, policies are created with respect to the (absolute) joint angle trajectories θ and relative to all $m \in M$ objects $\boldsymbol{o}^{(m)}$ detected in the environment. More precisely, the relative position of the robot's end effector to the initial position of each object (in Cartesian coordinates) is calculated.

In the reproduction phase, a Cartesian trajectory estimation $\hat{\boldsymbol{x}}^{(m)}$ and a related covariance matrix $\hat{\boldsymbol{\Sigma}}^{x(m)}$ for each detected object[1] is calculated by the Gaussian Process policies, giving M constraints. These constraints are used together with the joint angle estimation $\hat{\boldsymbol{\theta}}$ and related covariance matrix $\hat{\boldsymbol{\Sigma}}^{\theta}$ to create a final policy which considers constraints in both, Cartesian space and joint space. The projection from Cartesian space to joint space is done with a pseudoinverse Jacobian algorithm that compensates the missing orientation

---

[1] It is no problem if some of the objects from the training phase are not detected. Constraints with respect to these objects will be skipped.

information of the end effector [Lie77, CB08]. This can be accomplished through an optimization of the Jacobian null space with respect to the demonstrated joint angle trajectories (stay as close as possible to $\hat{\theta}$). The projection is applied to both the trajectory estimations $\hat{x}^{(m)}$ and the covariance matrices $\hat{\Sigma}^{x(m)}$ giving m additional joint space estimates $\hat{\theta}^{(m)}$, $\hat{\Sigma}^{(m)}$. To retrieving a generalized version of the trajectories, two distributions $\mathcal{N}(\hat{\theta}^{(i)}, \hat{\Sigma}^{(i)})$ and $\mathcal{N}(\hat{\theta}^{(j)}, \hat{\Sigma}^{(j)})$ are assumed as independent and the Gaussian product property is used as $\mathcal{N}(\theta, \Sigma) = \mathcal{N}(\hat{\theta}^{(i)}, \hat{\Sigma}^{(i)}) \cdot \mathcal{N}(\hat{\theta}^{(j)}, \hat{\Sigma}^{(j)})$. It is not necessary to perform the Gaussian product at each time step of the trajectory, because the Gaussian Process already delivers the covariances between all data points of the whole trajectory. All steps described here follow the procedure from [CB08, Cal07] with the option to consider constraints in *joint space only, task space only* or *joint and task space*.

Fig. 1 shows the demonstration of a *grasping and emptying* task. The resulting policies generated by the GP based framework are shown in Fig. 2.

| (a) | (b) | (c) |

Fig. 1. Demonstration of the *grasping and emptying* task. The first subgoal is to move the robot arm in proper position in order to grasp the blue cup (a). Subsequently, the arm has to move slightly besides the trash (b) to empty the cup into the trash (c).

2.3   Conventional Path Planning

So far, we considered learning from demonstration. A totally different problem is given in a situation where the robot's actuator starting pose and the desired goal pose are already known in advance. Here, all we have to do is to find a path from the start to the goal while avoiding obstacles. Path planning is the search for a continuous sequence of actuator joint configurations between an initial start configuration and a final goal configuration under the compliance of certain constraints. The term *motion planning* refers to the same problem, but usually describes a path parametrized by time (i.e. a trajectory).

There are several algorithms realizing path or motion planning by heuristic search algorithms. One example are *Rapidly-exploring Random Trees* (*RRT-Connect*) where the search space is explored from both directions, the start

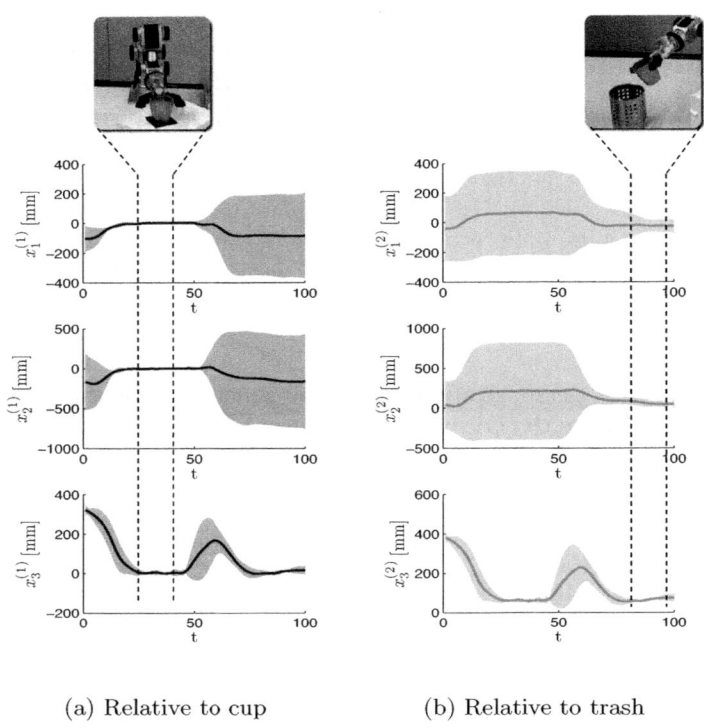

(a) Relative to cup   (b) Relative to trash

Fig. 2. The learned policies from a *grasping and emptying* task (similar to Fig. 1) relative to the initial positions of the cup (a) and the trash (b) in Cartesian coordinates. We can see that the movement relative to the cup is highly constrained between time steps 20 and 40. At this time, the end effector reaches the cup and grasps it. In contrast, the movement relative to the trash is constrained at the end of the trajectory (time steps 85 to 100), when the arm is able to unload the cup.

and the goal configuration, moving towards each other using a greedy heuristic [KSL00]. We use this algorithm due to its speed and accuracy. It is available within OpenRAVE, a powerful planning environment for robotics, supporting most known robot arms and also providing visualization, simulation and scripting interfaces [DK08]. Furthermore, OpenRAVE offers the possibility to search for paths not only avoiding obstacles but also considering arbitrary, self-programmed path constraints. For example, we can place a control function in the planner which will be called on every search step. This function will be allowing an explored joint configuration only if the end effector keeps a horizontal position, achieving a trajectory suitable to move a cup filled with a liquid. OpenRAVE also includes inverse kinematic solvers, offering the calculation of possible joint configurations for given end effector poses.

2.4   Combining Both Methods

An LfD oriented, GP based motion planning approach, as it was introduced in 2.2, and the conventional approach described in 2.3 cannot be compared directly because they address different problems. In the LfD framework, we have to extract constraints from demonstrations. Then, at the starting position when applying a skill, the problem is to generate a trajectory considering these constraints in a new situation. Regarding classical path planning, at the starting position we already know the initial and final joint configurations and the problem is to find a trajectory avoiding obstacles. What the methods have in common is the type of result, they both generate trajectories.

A trajectory from the LfD framework was built considering variances related to objects being present during demonstrations, but ignoring new objects in new situations. On the other hand, a classically generated trajectory is only considering the current state, but with all objects in the scene. It is obvious that the strengths can be combined.

Taking a closer look on the motion plan from the example in Fig. 3, we can see that there are low constrained regions (large variance) at the beginning and between time steps 60 and 80 regarding both objects (cup and trash). They refer to the first motion towards the cup and the motion towards the trash after the cup was grabbed. These are small regions regarding time, but they refer to the

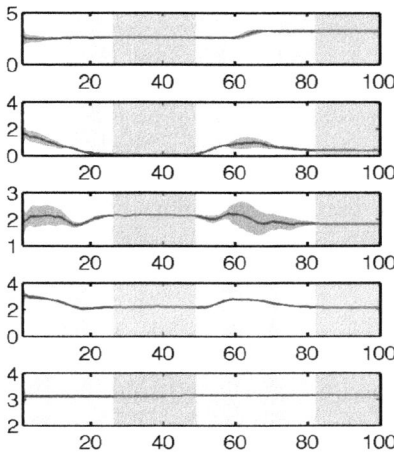

Fig. 3. The overall policy plotted for each DOF against time. This is the result of merging policies related to individual objects (as shown in Fig. 2). In the merging process, the smallest variances have the highest weights. Regions with low variance in each DOF independently are marked red. Green denotes the intersection of low variance regions for all DOFs. We cut out all non-green segments and generate new trajectories there with OpenRAVE. Highly constrained regions (green) remain as they are, since they represent the compliance of constraints learned from demonstration.

largest spatial motions. Hence, they represent the regions of highest collision risk regarding unknown objects. This is not an interpretation of this special example only. It is typical that we have large variances at regions of large spatial motions and small variances for motions of less spatial characteristics. A region of large variance means that it is low constrained and that we could move on many different paths to fulfill the required needs in the concerned segment. This is the most important property exploited in our new approach. Receiving a trajectory from the LfD framework, we cut out the regions of high variance with the intention to plan a new trajectory for these segments. Since we know the joint configurations at both ends of such a segment, we simply use classical path search avoiding all obstacles within the scene. The result is a powerful LfD framework, handling constraints from demonstrations as well as unknown objects in new situations.

At some paths, when low constrained regions are reached, a GP based motion still keeps constraints like holding a cup in a horizontal position due to the fact that the path is averaged over the demonstrations. Cutting the path at such a position and handing over responsibility to OpenRAVE could cause a subsequent plan not taking care of the end effector's angles, especially when obstacles have to be avoided. Therefore, at the end of a highly constrained path, the end effector's pitch and roll angles are read and a path planning constraint function in OpenRAVE is initialized, effecting OpenRAVE to keep up the mentioned angles.

3   Results and Discussion

We tested the method learning *grasping and emptying* and a simple *pick and place* task, each demonstrated four times on a *Neuronics Katana* arm. In situations where the learned skill had to be applied to, all known objects have been repositioned and one or two new objects have been added to the scene. The method performed well reproducing demonstrated skills considering new objects successfully avoiding them as obstacles. When moving the cup towards the trash, it was successfully held in a horizontal position in path segments generated by OpenRAVE (cf. Fig. 4).

Problems can occur if new objects are placed very close to those being manipulated during demonstration. In such cases, the first position of a segment with a high variance and thus the starting position for the classical path plan done by OpenRAVE can already be a position where the robot arm collides with an object. Hence, OpenRAVE will not be able to find a path. Nevertheless, this is still a correct result. In such a case, it is simply not possible to reproduce the skill without a collision in the manner how it was demonstrated. Then, this is not a problem the low-level framework should deal with. A consequence might be to delegate the problem to a higher, symbolic planning level, which, for example, could decide that the obstacle should be moved aside first.

Naturally, there is still room for improvement. For example, the system could determine automatically, whether the end effector angle constraints should be handed over to OpenRAVE, as it is now done for every case.

(a)   (b)   (c)

Fig. 4. Reproduction of the *grasping and emptying* task. First with the known objects only (a), then with an unknown obstacle (one trajectory split up into images b and c). The paths generated by GP (red) and by OpenRAVE (green) are marked. The situation in (a) can be solved by using only GP. The new situation with the obstacle can only be solved by the extended new framework.

In further experiments, it has to be investigated whether the technique to cut out low constrained regions is applicable to a wide range of problem classes. In these segments handed over to OpenRAVE there are still low variance regions regarding single DOFs. Usually, segments being low constrained for a single DOF only result from a lack of variation during demonstration. Nevertheless, maybe there are some problem classes where this does not apply to.

Another open task is the interaction with a higher level symbolic planner as already mentioned above or the integration into a classical three-tier robotic architecture.

4   Conclusion

We showed that a GP based LfD framework combined with conventional path planning methods can enhance the generalization capabilities of a robot regarding learned skills. Thereby, strengths from both approaches are exploited. The class of unknown situations the robot might face to apply learned skills was expanded successfully by objects and obstacles which are not known from demonstration.

5   Future Work

Additional benefits could be derived from the usage of a robotics learning framework. Instead of relying on a single learning technique to plan motions, learning results could be improved by combining different methods. This could, for instance, be a combination of Gaussian Processes and Human Trainer Feedback or Reinforcement Learning.

As a toolbox which provides the functionality to switch between multiple learning and planning algorithms and to combine them, the TEACHINGBOX

[ESCT09] could be used to reach this goal. Prospectively, the current motion planning algorithm is going to be integrated into the TEACHINGBOX framework. Regarding real-world examples like the Katana robot arm, appropriate use cases will have to be developed in this case.

Acknowledgement. This work was supported by the Collaborative Center of Applied Research on Service Robotics (ZAFH Servicerobotik).

References

[BCDS08] Billard, A., Calinon, S., Dillmann, R., Schaal, S.: Robot programming by demonstration. In: Siciliano, B., Khatib, O. (eds.) Handbook of Robotics. Springer, Heidelberg (2008) (in press)

[Bug03]   Bugmann, G.: Challenges in verbal instruction of domestic robots. In: Proceeding of the ASER 2003, The 1st International Workshop on Advances in Service Robotics (2003)

[Cal07]   Calinon, S.: Continuous Extraction of Task Constraints in a Robot Programming by Demonstration Framework. PhD thesis, Learning Algorithms and Systems Laboratory (LASA), Ecole Polytechnique Federale de Lausanne, EPFL (2007)

[CB08]   Calinon, S., Billard, A.: A probabilistic programming by demonstration framework handling skill constraints in joint space and task space. In: Proc. IEEE/RSJ Intl. Conf. on Intelligent Robots and Systems (IROS) (September 2008)

[DK08]   Diankov, R., Kuffner, J.: Openrave: A planning architecture for autonomous robotics. Technical Report CMU-RI-TR-08-34, Robotics Institute (July 2008)

[EK08]   Ekvall, S., Kragic, D.: Robot learning from demonstration: A task-level planning approach. International Journal on Advanced Robotics Systems 5(3), 223–234 (2008)

[ESCT09] Ertel, W., Schneider, M., Cubek, R., Tokic, M.: The Teaching-Box: A Universal Robot Learning Framework. In: Proceedings of the 14th International Conference on Advanced Robotics (ICAR 2009), Munich (2009), http://www.teachingbox.org (accessed February 25, 2010)

[EZRD02] Ehrenmann, M., Zoellner, R., Rogalla, O., Dillmann, R.: Programming service tasks in household environments by human demonstration. In: IEEE Intl. Workshop on Robot and Human Interactive Communication, pp. 460–467 (2002)

[GWB97]   Goldberg, P.W., Williams, C.K.I., Bishop, C.M.: Regression with input-dependent noise: A gaussian process treatment. In: Neural Information Processing Systems. MIT Press, Cambridge (1997)

[KMP08]   Kober, J., Mohler, B.J., Peters, J.: Learning perceptual coupling for motor primitives. In: Proceedings of the 2008 IEEE/RSJ International Conference on Intelligent Robots and Systems (IROS), pp. 834–839. IEEE, Los Alamitos (2008)

[KPPB07] Kersting, K., Plagemann, C., Pfaff, P., Burgard, W.: Most likely heteroscedastic gaussian process regression. In: Proceedings of the Twenty-Fourth International Conference on Machine Learning, ICML 2007, Corvalis, Oregon, USA, June 20-24. ACM International Conference Proceeding Series, vol. 227, pp. 393–400 (2007)

[KSL00]   Kuffner Jr., J.J., Lavalle, S.M.: RRT-connect: An efficient approach to single-query path planning (November 06, 2000)

[Lie77]   Liegois, A.: Automatic supervisory control of the configuration and behavior of multibody mechanisms. IEEE Trans. on Systems, Man and Cybernetics SMC-7(12), 868–871 (1977)

[NW99]   Nocedal, J., Wright, S.J.: Numerical Optimization. Springer, New York (1999)

[PS07]   Peters, J., Schaal, S.: Policy learning for motor skills. In: Ishikawa, M., Doya, K., Miyamoto, H., Yamakawa, T. (eds.) ICONIP 2007, Part II. LNCS, vol. 4985, pp. 233–242. Springer, Heidelberg (2008)

[RW06]   Rasmussen, C.E., Williams, C.: Gaussian Processes for Machine Learning. MIT Press, Cambridge (2006)

Development, Realization and Control
of a Mobile Robot

Ralf Stetter, Paweł Ziemniak, and Andreas Paczynski

Hochschule Ravensburg-Weingarten, Doggenriedstraße,
88250 Weingarten, Germany

Abstract. This paper reports the development, realization and control of a mobile robot which mainly aims at applications in the area of production logistics. The unique characteristic of this mobile robot is its steering principle. This patented steering principle is based on the usage of torque differences between individually driven wheels, which can align into the direction of the desired robot path. This steering principle leads to excellent maneuverability but requires an elaborate control system. The realization of this control system is the main contribution of this article.

Keywords: Mobile robots, autonomous vehicles, control systems.

1 Introduction

Robotics is a field that fully exploits the term mechatronics: advanced mechanical construction, smart sensors, smart actuators, medium to large number of processing units, and sophisticated software system. This publication describes a vehicle with the distinct quality that it could not drive at all without a functioning control system. Due to the unique steering principle a coordination of different drive units is needed in order to align these drive units. Therefore, this robot can be considered a high-level mechatronic system which main functions are depended on all three disciplines of mechatronics. In the Systems Engineering Laboratory of the Hochschule Ravensburg-Weingarten in Weingarten, Germany mobile robots are developed and thus an effective and efficient methodology is necessary in order to achieve the set goals. It was found that as general basis the V-model is appropriate to manage all designs in this laboratory. Besides the V-model, several other methodologies and supporting tools are applied in order to be able to continuously improve the designs under development. Stania&Stetter [1] propose to refer to the whole process also integrating organizational aspects as "Mechatronics Engineering".

2 Development of the Production Vehicle

A few years ago an innovative steering system for vehicles of any kind was developed and patented at the Hochschule Ravensburg-Weingarten [2], [3]. In order to validate this concept, a production vehicle was developed and realized which is used to display the full potential of the innovative steering concept as well as the general

D. Obdržálek and A. Gottscheber (Eds.): EUROBOT 2010, CCIS 156, pp. 130–140, 2011.
© Springer-Verlag Berlin Heidelberg 2011

applicability in industrial surroundings. This vehicle was designed in order to meet the following central requirements:

- the ability to transport materials of a maximum mass of m=500kg (large transmissions, automotive engines, etc.);
- a minimal vehicle velocity of v = 1m/s;
- a minimal vehicle acceleration a = 0.5m/s^2
- overall dimensions of the vehicle 1200[cm] x 800 [cm] (according to a Euro palette);
- excellent maneuverability and high dynamics;
- autonomous driving;
- additional functions (e.g. scalability).

These requirements were gathered in cooperation with a regional company which develops and produces unmanned guided vehicles for production logistics. Figure 1 shows an example of the production vehicle.

Fig. 1. Production vehicle (CAD-model)

Extensive investigations in the development of this innovative vehicle led to a completely modular design. Special attentions deserve the driving units of the autonomous industrial vehicle. Thanks to the innovative and patented solution of the steering system of the vehicle the possible movements, capacity, efficiency, speed and the robustness could be improved in comparison to other more conventional production vehicles. This steering system is based on a concept which uses torque differences between the drive motors of the wheels to steer the angular position of each module. Figure 2 shows the steering principle: if one module (consisting of two wheels each driven by its own motor) is not in the angular position which is required for the movement of the whole vehicle then the desired angular position (orientation) can be achieved through sensible control of the torques of the two drive motors.

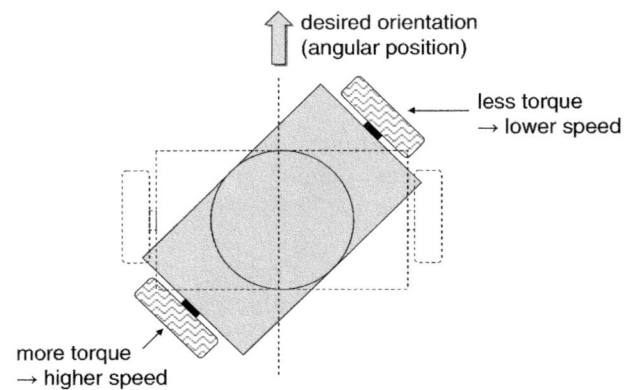

Fig. 2. Torque steering in one module

This general concept was realized as an intelligent driving unit called module. This subsystem consists of two electrical motors units (called compact drive) MCD EPOS, one of them is programmable - P (Programmable) and acts as master, the second one is not programmable and acts as slave - S (Slave). The compact drives dispose of an encoder, electric motor and controller. Already such compact drive is an intelligent actuator in itself. Figure 3 presents the detail design of one module.

1 - encoder, 2 - brake, 3 - gearbox, 4 - horizontal axis, 5 - vertical axis, 6 - bearing, 7 - motor and controller with gear head, 8 - slip-ring

Fig. 3. Detail design of a drive module (cross-section – CAD model)

The mobile robot uses two kinds of data buses and three power supply levels. The main bus is a CAN bus which interconnects all the modules with the industrial PC and multimeter. The other bus is I2C which interconnects all the ultrasonic sensors and optional compass with the industrial PC. The 24V power line supplies all the modules, the 12V supplies the industrial PC and the 5V is required for the ultrasonic sensors. The 12V line comes primary via DC/DC converter from 24V batteries but when the voltage drops below threshold the line will be supplied by backup 12V battery. This way the computer is safe from failures that originate from voltage drops caused by the motors. In future, however, the PC can be replaced by one appropriate for 24V system allowing dropping the 12V line but requiring additional securing. The physical connections are shown in Figure 4.

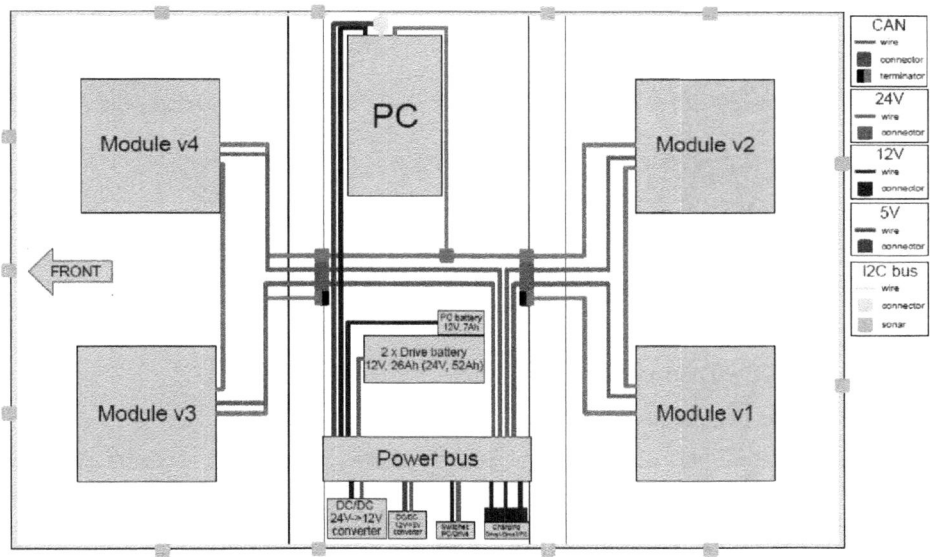

Fig. 4. Physical connections in the mobile robot

3 Control System

This section describes firstly the control system architecture and then the control system structure. The heart of the control system architecture is an industrial PC which interconnects all the devices by means of various interfaces (CAN, USB, I2C, Ethernet). The devices within the modules (motor controllers and encoders) are connected to the CAN bus, the brakes are connected to digital outputs of the controllers. The graph in Figure 5 presents the architecture of the control system; the connections with solid lines are permanent and required for normal operation and the other connections are optional in order to control the robot locally or remotely.

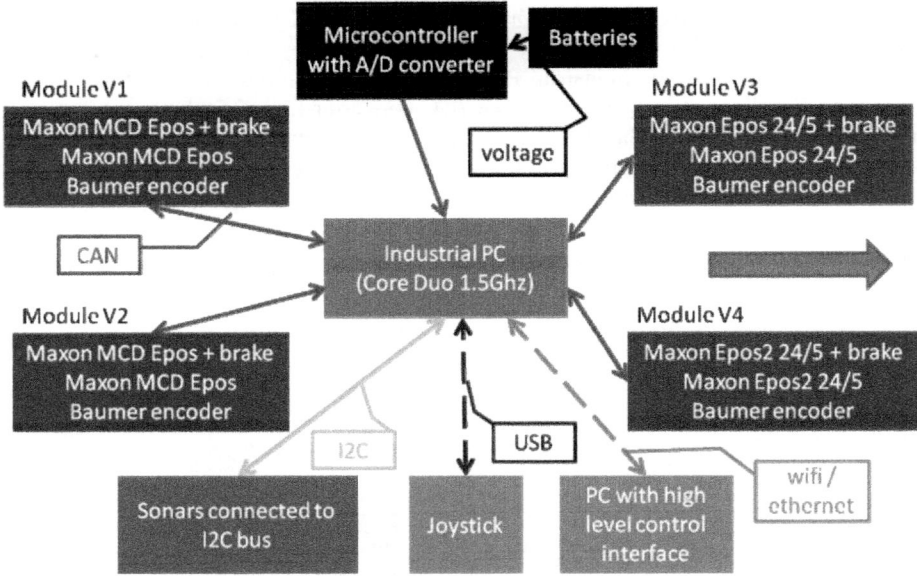

Fig. 5. Control system architecture

The control system consists of three levels:

- The high level contains:
  - the control interface which analyses the joystick movement, sends TCP packets with instructions to the control system as well as asks and receives information about the robot status (batteries, sensors, velocities, currents, etc...),
  - the autonomous operation module which sends movement instructions according to predefined trajectory, navigates using ultrasonic sensors and odometry, performs simple localization and odometry error elimination, and communicates via TCP with the control system (application preferred to run on local host (control system PC) but can also operate remotely),
  - the system analysis module which is realized by means of a set of matlab scripts which run on the remote machine and connect via TCP; it visualizes the system including module angles, position in global coordinate frame, analyzes the motor velocities and currents and analyses the battery levels during operation/charging states.
- The middle level (control system):
  - is a real-time priority application on the robot PC,
  - listens for TCP packets with commands, interprets commands, distinguishes operation modes (driving modes) and driving parameters (typical drive instruction: mode + angle + velocity)

- contains the kinematics model which recalculates the demanded movement angle (trajectory) to angle for each of the modules and recalculates the wheel velocities according to demanded robot velocity and curvature,
- applies some profiling regarding robot acceleration,
- uses separate PID to control the angle of each module,
- applies additional profiling regarding velocity of each motor, sends demanded velocities to the motor controllers,
- gathers all sensor information and provides to clients when asked (distances, velocities, currents, etc..) and
- analyzes information from ultrasonic sensors and slow-downs or stops the robot when proximity thresholds are reached.

- The low level (motor controllers)

  - where a pure velocity mode is used to ensure appropriate velocity convergence;
  - velocity profiling is done already on the higher level.

The control system was written in C++ language using the Visual Studio C++ 2008 environment. The selection of language was initiated by requirements of such software – mostly the necessary speed of control system. The environment may change in case of porting the code into a Linux machine which was considered later as possible improvement for further robotics projects.

A part of the UML class diagram of the control system is shown in Figure 6 (intentionally clipped section). The system consists of classes which names correspond to hardware devices of the robot or some functionalities. The first set of classes presents the robot main class IndustrialRobot. The class is the heart of the system and contains most of the functionalities of the control system. The class has three derivatives which correspond to various robot configurations (1 module, 3 modules, 4 modules). The derived classes implement only the appropriate kinematics model. Various other derivatives can be easily provided with any number of modules. Some of the robot attributes visible in the figure are: DrivingParameters which contains the most current orders for the robot and SensorReading which contains pre-processed readings gathered from sensors. These attributes are thread safe (inheriting from TThreadParam). This is necessary since the control program is multithreaded and appropriate synchronization is required. The separate threads are responsible for TCP/IP communication, sensor readings, voltage measurements, control of driving modules and software watchdog which stops robot when problems occur (lost communication).

On another section of the diagram also a class called AccurateTimer could be found. The class is a singleton which easies access to high accuracy multimedia timer. The resolution of the timer is 1/1000 of a second what is enough for the project.

Figure 6 only shows about ten percent of the UML class diagram, many other classes are necessary in order to allow a control of the mobile robot.

One cornerstone of the control system of the mobile robot is the kinematic model. Figure 7 shows the kinematic model of the mobile robot.

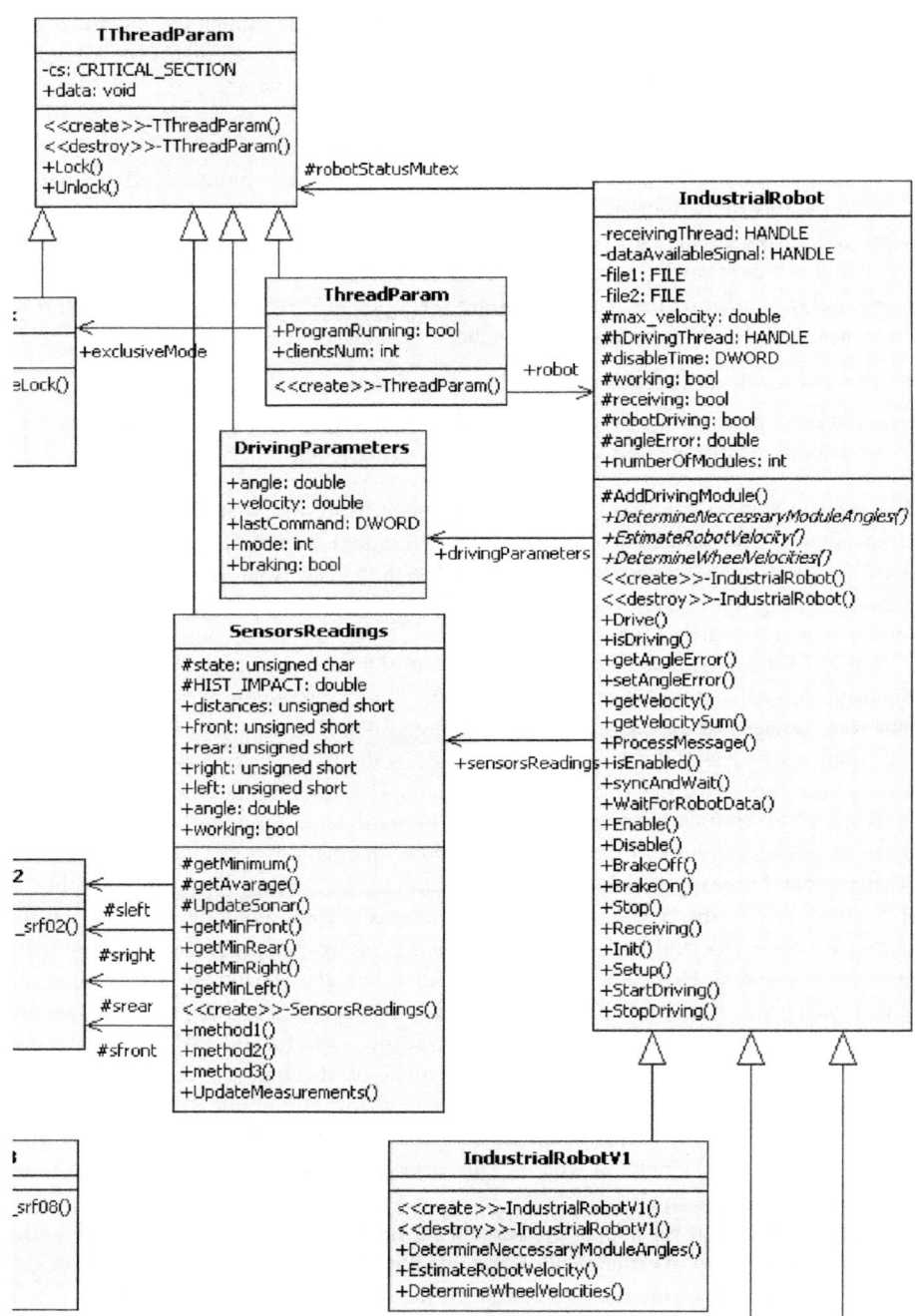

Fig. 6. Clipped section of the UML class diagram of the control system of the mobile robot

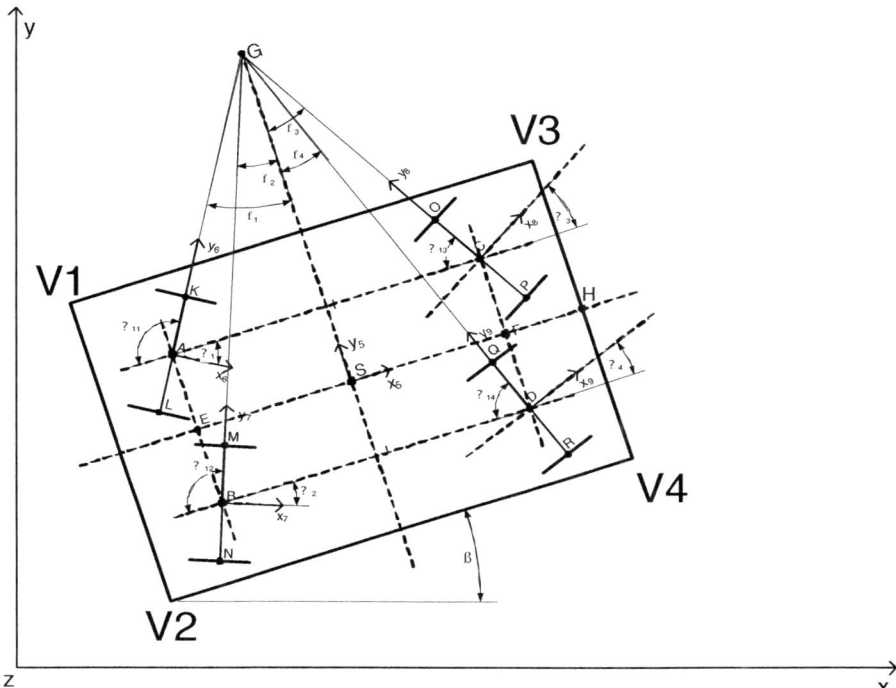

Fig. 7. Kinematic model of the mobile robot

The robot is placed in the global coordinate frame (x, y, z). The robot chassis coordinate frame is (x_5, y_5, z_5) and contains all the modules (V1, V2, V3, and V4) with the frames (x_6, y_6, z_6), (x_7, y_7, z_7), (x_8, y_8, z_8) and (x_9, y_9, z_9). The trajectory is given to point H which is in front of the vehicle. The point G denotes instantaneous centre of curvature (ICC) and is placed on the vertical symmetry line of the robot. Such placement of this point decreases the number of degrees of freedom. Putting this point elsewhere will increase number of possible manoeuvres but only in this mode all the modules are balanced – the absolute of turning angles of front and corresponding rear modules are equal. For movement along straight lines in any directions different model is used. This case the model is very simple since all the modules have equal angles and all the wheel velocities are also equal.

4   Realization and Autonomous Operation

The mobile robot was realized in the workshops of the Hochschule Ravensburg-Weingarten during numerous projects, bachelor theses and master theses. The robot uses industrial modules, were possible for the sake of easy realization but also for the sake of high reliability and easier adaption the industry conditions. Figure 8 shows the realized production vehicle (here with three modules – at this moment in time the fourth module was still in production).

Fig. 8. Realized mobile robot (with three modules)

After finishing the mechanical design and control system, the robot was used for research concerning autonomous operation. Unfortunately, due to high cost, there was no possibility to use a 3D laser scanner on the robot. Instead, a combination of infrared and ultrasonic sensors was proposed. Infrared sensors were supposed to measure exact distances while ultrasonic were used for the navigation. Due to characteristics of the production hall, finally, only ultrasonic sensors were used. Such setup was enough because ultrasonic sensors return obstacle proximity in much wider angle then infrared ones. The infrared could be used for fine tuning and more precise localization which was not so important at the beginning of the research.

The robot has 13 sensors located around it; the sensors are perpendicular to the edges and in equal distances within the edge. There are three sensors in front of the robot, two on the back and four on each of the sides. Three of the sensors, centre sensor on front and first sensor on left and right, are more precise measuring distances from 3cm to 6m (SRF08) while the others measure from 15cm to 6m (SRF02). The three sensors have also light detector which was not used in the setup.

For more precise localization also an I2C compass module was employed. Unfortunately the compact design of the robot prevented the use of compass – it needs to be distanced from the steel construction. Moreover, the compass would not work in a production hall, instead of north, it would point to the nearest machine and would get additional noise when the machines operate.

The final setup consisted only of 13 ultrasonic sensors and from odometry of all the wheels of the system. Each of the modules has two wheels what gives some redundancy and increases precision of odometry. When one wheel slips it is still possible to detect it and compute distance correctly. Thanks to this, the setup can be based mainly on the dead reckoning. Appropriate inverse kinematics model was synthesized for that purpose. The production hall was measured first using standard tools and then using robot to verify. As expected the robot odometry is not free of errors and processing the same route several times only increases the error. In order to improve accuracy it is necessary to perform localization form time to time to adjust position of the robot. Several places were found in production hall which could be used as landmarks. Some of the landmarks where also the machines from and to which robot had to deliver goods. When approaching the machine robot tried to detect precisely the environment around in order to place itself in demanded place for load/unload. Similar situation was in well known places that could be easily analyzed with available sensors. This way dead reckoning was improved and system had enough accuracy to perform the same loop multiple times.

The task of the robot was to deliver goods between four machines. The four machines and a corridor with specific configuration on left and right were used as landmarks. The robot could proceed the path repeatedly unattended. For safety reasons a hardware cut off button was available on the robot and software cut off button was available at configuration station. When the robot path was interrupted by an obstacle (human) the robot stopped waiting for clear path. Operation in the hall in presence of such obstacle would be dangerous due to limited space and sensors uncertainties. The stop-when-obstacle policy was enough for testing and gave good results.

Although the setup met the requirements, a lot of improvements could be done. First of all sensory system could be improved. The 3D laser could give very accurate view on the environment but the improvement can be also made by installing mentioned IR sensors, more ultrasonic sensors preferably in the ring or accelerometers that directly support dead reckoning. With more precise information the position accuracy and information about obstacles would be sufficient to operate in dynamic environments – omitting the moving obstacles. With even more logic and in bigger scenarios also different routes could be taken if obstacle decrease safety or even completely block some paths.

5 Summary

The development and realization of mobile robots an their control system presents a major challenge to engineers of the three disciplines mechanical engineering, electrical engineering and software engineering. This paper describes the successful realization of a mobile robot for production logistics at the Hochschule Ravensburg-Weingarten. One main challenge was the control system. This control system was described on two levels: firstly the control system architecture and secondly its structure. Due to the application of strategies, methods and tools, e.g. the application of a procedure roughly following the V-model or the use of the models of UML a promising prototype could be realized. The first feedback of the industrial partner was very positive.

References

1. Stania, M., Stetter, R.: Mechatronics Engineering on the Example of a Multipurpose Mobil Robot. In: Solid State Phenomena, vol. 147-149, pp. 61–66 (2009)
2. Stania, M., Stetter, R., Ziemniak, P., Paczynski, A.: Intelligentes Steuerungssystem für autonome Fahrzeuge in Service- und Produktionsanwendungen. In: Bericht über die Tagung Mechatronik 2009 – Komplexität beherrschen, Methoden und Lösungen aus der Praxis für die Praxis, S.101–S.108 (2009)
3. Stetter, R., Paczynski, A., Zając, M.: Methodical development of innovative robot drives. Journal of Mechanical Engineering (Strojniski vestnik) 54, S.486–S.498 (2008)

Startup Robotics Course for Elementary School

Dmitry Sukhotskiy and Anton Yudin

Moscow Information Analytical Center (IAC), Robotics laboratory,
1-y Zborovkiy pereulok, 3, 107076, Moscow, Russia
Bauman Moscow State Technical University, IU4 Department,
2-nd Baumanskaya st., 5, 105005, Moscow, Russia
{Dsuhotsk,skycluster}@gmail.com

Abstract. The article is an example of preparatory course for studying robotics by junior school children and aims at forming an educational framework for future expansion of the course. Main topics for studying were selected on the basis of author's experience at elementary school age. Recommended curriculum and examples which are illustrative and easy to understand at the junior age are also discussed.

Keywords: Eurobot, robot, education, programming, elementary school.

1 Introduction

Author experience includes active participation in scientific and technical exhibitions and conferences and robot development for Eurobot [1] and Eurobot Junior competitions for more than 3 years. When grownups see the work of a kid hardly out of elementary school they usually ask one and the same question: "How did you start? What is the right way to start teaching robotics to younger children?"

At initial phase of studies with children it is important to form the right conception of how basic robot components (programs, electronic parts and actuating mechanisms) interact with each other. At the end of this phase a child will know the ultimate purpose of work, the concept of a robot. As a result wish and motivation to move on with robotic studies will arise.

Based upon author's experience an educational programme is developed which includes logic elements explanation, visual programming and development of a simple robot, controlled with a personal computer (PC).

With assistance of Russian Eurobot NOC [2] a series of seminars were held with a group of pupils who afterwards participated in Eurobot Junior 2010 competitions. It made possible to validate and update study topics of the course and to form a list of necessary equipment and parts for it.

2 How Does Technical Knowledge Form in Children?

Child's early awareness of electrical engineering elements and electronics is driven by modern life conditions. Children are facing electronic devices (lighting and heating

D. Obdržálek and A. Gottscheber (Eds.): EUROBOT 2010, CCIS 156, pp. 141–148, 2011.

devices, cooking machines and other consumer electronics) at extremely young age. Toys with electronic parts usually trigger initial interest to look inside and to understand "how it works." By doing so, a child is getting basic knowledge of electric circuits and associated parts: current sources, lamps, motors, switches, etc. It is possible that from such toys the interest to robotics arises.

A major role in studying electronics play different sorts of construction kits. In Russia there is a kit named "Znatok" [3]. Easy-to-use electrical connections and a lot of circuit designs of different complexity make it unique. One may use it in learning process for as long as 2 or 3 years.

Amount of useful information in a toy is limited and fast to comprehend. But nowadays kids have a rather permanent "toy" – computer. There is even a term – computer kids. The answer to a question of usefulness of computers highly depends on the purposes the computer is used for. In robotics studies PC is an irreplaceable assistant. From early age children may draw, type text and understand that computers control such devices like printer or scanner. It is enough to give child and idea that same computer can control any simple device, e.g. a toy. And the aim of the proposed course is to teach them how.

3   Study Topics

There are no exact seminar in-depth descriptions in this article. Instead, several most important recommended subjects and examples are presented. They are intuitive and easy-to-understand for kids.

Conscious interest in knowledge children show only while in game. That is why it is important to conduct only practical seminars in game style. First part of such seminar should be dedicated to circuit assembly and the second – to programming. Circuits are assembled on a bread board without any soldering.

It is necessary to include these topics in curriculum:

- Logic elements;
- Visual programming environment Borland C++ Builder;
- Proteus Design Suite – for modeling electronic circuits;
- Motor driver L293D;
- Resulting robot assembly, able to be operated with computer.

3.1   Logic Elements

Logic element is an electronic device in which any state of input signal is associated with a definite state of the output signal. It is not necessary to assemble all possible circuits of basic logic elements on a bread board. Any of the circuits gives a straight idea of how logic works. But truth tables should be thoroughly reviewed.

First seminar should be started with a simple AND element and its 2-switch representation. Circuit diagram, symbol and truth table for this element are presented below. Logical "one" is produced by the element only when both inputs are logical "ones". This explains the name of this element as "ones" should be present on one AND the other (both) inputs.

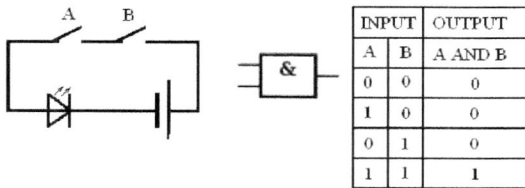

INPUT		OUTPUT
A	B	A AND B
0	0	0
1	0	0
0	1	0
1	1	1

Fig. 1. Circuit diagram, symbol and truth table for AND element

When explaining what a composed element is a simple NAND element can be used. Its work is similar to AND, but output signal is opposite to the input. When AND has logical "one" on output, NAND has logical "zero" and vice versa. This is easy to understand with the help of the figure below.

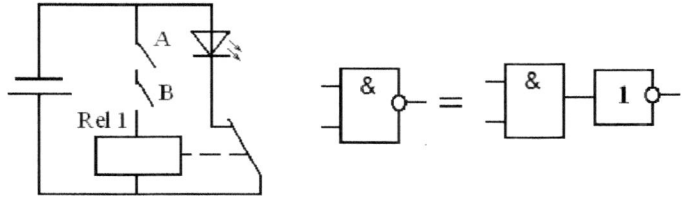

Fig. 2. Circuit diagram and equivalent symbol of NAND element

Combination of logic elements is demonstrable with SR flip-flops based on two logic elements.

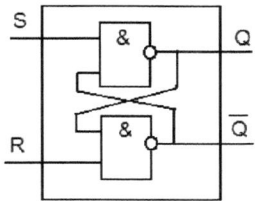

Fig. 3. SR flip-flop

Flip-flop is a logic element that can be for a long time stay turned on and cycle its states according to external signals. Here it is important to revise truth tables of NAND elements once again and compare to that of a flip-flop.

Assembly of simple circuits of encoder and decoder can help to make the first step in learning binary arithmetic. Encoder translates single input signal to binary code on output.

In the next figure one can see an encoder based on 4 OR logic elements. Diagram shows an input signal for decimal "5" number. Diagram visualizes that when logic one is introduced to input dedicated to decimal "5" first and forth outputs will generate logic "ones", thus forming 0-1-0-1 binary number.

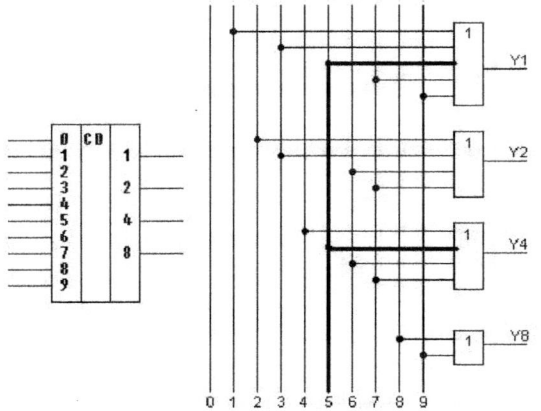

DECIMAL	Y8	Y4	Y2	Y1
0	0	0	0	0
1	0	0	0	1
2	0	0	1	0
3	0	0	1	1
4	0	1	0	0
5	**0**	**1**	**0**	**1**
6	0	1	1	0
7	0	1	1	1
8	1	0	0	0
9	1	0	0	1

Fig. 4. Encoder

For back translation of binary to decimal numbers special decoders are used. When a binary number is on the input of the decoder a dedicated single output is activated.

In figure 5 one can see the symbol and truth table of the 2-to-4 decoder. During the seminars it is sufficient to look only into the truth table.

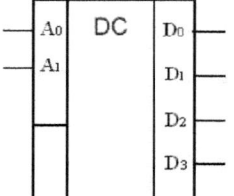

INPUT		OUTPUT			
A_1	A_0	D_3	D_2	D_1	D_0
0	0	0	0	0	1
0	1	0	0	1	0
1	0	0	1	0	0
1	1	1	0	0	0

Fig. 5. Decoder

It is important to note that decoder can be used to decrease the number of connections between electronic components; in our case it is possible to control 4 devices one by one. For real applications it is better to use a 4-to-16 decoder.

3.2 DLP-USB245M, UM245R Modules

During the course seminars DLP-USB245M or UM245R module was used for interfacing computer to peripheral devices. These are converters of USB signal to simple parallel interface. One doesn't need to know USB protocol to operate these modules but it is required to download and install module drivers and libraries. Each module has a number of useful functions and modes. For educational purposes only one is sufficient – connect all 8 channels of the module as outputs. In such mode it can be compared to previously discussed encoder. It should be called repeater though as it repeats on module outputs the number submitted to computer program.

3.3 IDE Borland C++ Builder Programming

First programming lessons should be more visual, for example one could use Borland C++ Builder environment. Several lessons will be enough for children to learn a number of primitives and how to use them in a program. These could include buttons, text editor, progress bar, etc.

For startup projects while on initial stage of robotic study it is better to use previously prepared and checked program code, which is necessary to change according to current design needs.

There are a lot of literature and Internet sources with program examples for self-instruction of how to work in a programming environment. Below some code examples are presented to describe how control of periphery devices through DLP-USB245M module in Borland C++ Builder environment could be achieved.

Module start.

```
FT_HANDLE ftHandle;
FT_STATUS ftStatus;

ftStatus = FT_Open(0,&ftHandle); //Open port and set
identifier 0, which means one device in operation

if(ft_Status == FT_OK) {
    } else Caption = "Open Failed";
```

Next step is activating all channels to output data. When program is running function FT_SetBitMode() sets signal lines D0..D7 to output "1" or "0".

```
FT_SetBitMode(ftHandle,0xFF,1);
```

Now all output lines are activated to output a hexadecimal number FF, which means 11111111 in binary code.

Set of "high" level on D1 output with button 1.

```
Void __fastcall TForm1::Button1Click(TObject *Sender) {
    State = 2; // set variable
    ftStatus = FT_Write(ftHandle, &State, 1,
&lpdwBytesWritten);}
```

Print current state of outputs on the screen.

```
FT_GetBitMode(ftHandle, &BitM);
Label1->Caption = BitM;
```

For convenience control buttons can be presented on screen as arrows indicating the motor turn direction or movement back and forth.

Pressing a mouse button generates sending a decimal number to port, "4" for example. Module output lines generate a binary number "00000100". As a result "high" voltage level is introduced to third line and it can be used to control motor driver input or some other device.

3.4 Motor Driver L293D

There are a number of different designs to control electric motors. They differ in power and hardware components. Motor driver L293D [4] – is a unique design used by each and every beginning robot technician. It works similarly to a logic element.

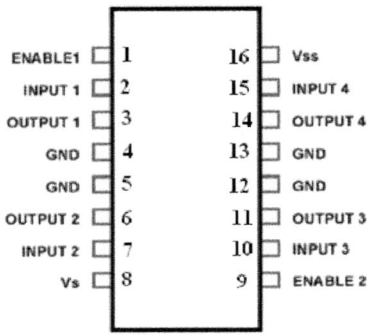

Fig. 6. Motor driver L293D

During seminars it is important to highlight that motors are controlled bidirectional. Second important point is ENABLE inputs of the driver. For better understanding table below is presented.

Table 1.

ENABLE1	INPUT1	OUTPUT1
1	0	0
1	1	1
0	0	0
0	1	0

3.5 ISIS Proteus

For modeling electronic circuits, there are a lot of popular programs. During this course, ISIS Proteus Design Suite [7] was used. To master it within a short period of time is not possible, but it is important to give a general idea. Children quickly memorized how to draw circuits and how to find familiar components in a library. With

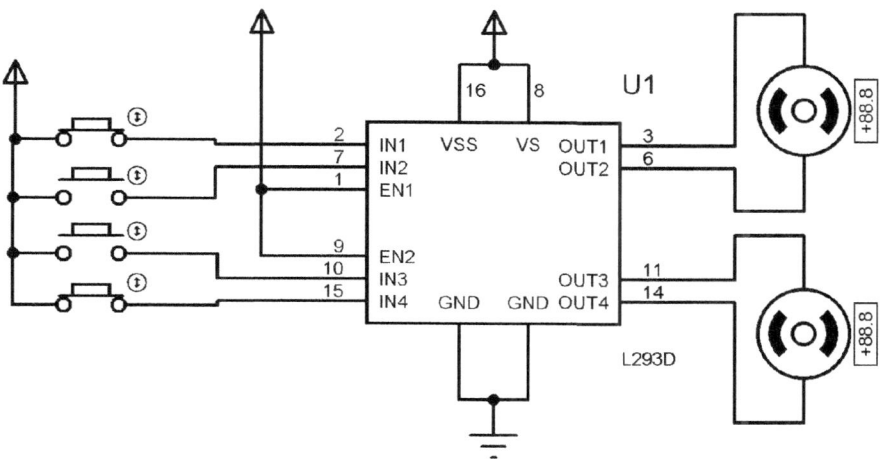

Fig. 7. Motor connection diagram

great relish they drew diagrams, putting virtual LEDs and motors into action. The figure below presents motor connection diagram to L293D driver.

4   Results of Conducting a Startup Robotic Course for Pupils

Author was not limited in choosing topics for seminars or methods of teaching. But wishes of kids in a group and their parents were always taken into consideration. This way was conducted one of the seminars when a number of different relays and motors were disassembled. Soldering lesson was also conducted. Using bread board and thin conductors is rather diligent work and quickly annoys children. As a consequence the most number of circuit designs was realized on computer.

On the last seminar a group was divided into two teams. Each team assembled its robot during the seminar and in the end a competition was held. After the course was over, all participants decided to make a robot for Eurobot Junior. Some of them involved their parents in robot construction and made their own robots at home.

As a result of this first and experimental robotics course among junior school children valuable teaching experience was gained. A portion of this experience was presented in this article. Authors believe to contribute more to the course in future and form an educational framework for teaching basics of robotics according to gained experience. Such framework is to be described in future articles.

During the seminars work of students inspired an idea of a simple robotic construction kit. Several beta-versions of parts of such a kit were realized and used throughout the course. It was decided to develop a simple robotic construction kit for future startup robotic courses, which would allow assembling circuits without soldering, including a bread board, electronic components, motors and control software. The kit will be designed bearing in mind that many parents have little experience in electronics and programming but still wish to practice with their children at home.

References

1. Eurobot, international robotics contest, http://www.eurobot.org
2. Eurobot National Organizing Committee (NOC) of Russia, http://www.eurobot-russia.ru
3. Znatok electronic construction kit, http://www.znatokonline.ru
4. Datasheet for L293D Motor Driver, http://www.st.com/stonline/products/literature/ds/1330.pdf
5. Logic elements at Wikipedia, http://en.wikipedia.org/wiki/Logic_gate
6. UM245R Drivers, program examples, datasheets, http://www.ftdichip.com
7. Proteus Design Software official website, http://www.labcenter.co.uk/products/pcb_overview.cfm

Micromouse – Electronics on Wheels

Tony Wilcox

Birmingham City University
Faculty of Technology, Engineering and the Environment
Millennium Point, Birmingham B4 7XG
tony.wilcox@bcu.ac.uk

Abstract. Micromouse at TEE[1], evolved as a mechanism to enthuse and engage students more fully in the practical aspects of designing real-time embedded systems, from both hardware and software perspectives. Static target boards are ideal for an introduction to embedded systems - to learn the basics of embedded hardware and software. Mobile robots bring in motor drives, positional encoders, distance sensors, closed-loop control theory, real-time systems design, mechanical considerations, and real-world problems relating to the multivariate environments in which these devices must operate. Mobile Robots are "Electronics on Wheels" – experience shows that things that move are far more engaging than things that don't.

The design, build, and programming of a micromouse requires multi-disciplinary skills. It provides the opportunity for individual students to develop broad technical capability, for groups of mechanical, electronic, software and embedded systems students to develop team-working skills, and of course it has the element of competition that makes it fun and rewarding.

Birmingham City University hosts the UK Micromouse competition, as well as several smaller events throughout the year. UK Micromouse is the major event of BCU TECHFEST[2]. The range of competitions has been extended to incorporate Schools Micromouse, which brings in line-followers, drag-racers and wall-followers as well as maze solvers. The aim is to support the STEM agenda – encouraging the study of Science, Technology, Engineering and Mathematics.

This paper discusses the design of a micromouse for teaching embedded systems. In addition a model for micromouse software development is presented.

1 MyTEEmouse

The earliest micromouse designed at BCU was Robotic (Fig.1 and Fig.2), designed originally as a student project[3] which was used in its various forms for 6 years from

---

[1] TEE refers to the Faculty of Technology, Engineering and the Environment of Birmingham City University, Birmingham, UK.
[2] BCU TECHFEST is an annual event hosted by Birmingham City University, Faculty of Technology, Engineering and the Environment, TEE, held at Millennium Point, Birmingham, B4 7XG.
[3] S.Sample, University of Central England, UK.

D. Obdržálek and A. Gottscheber (Eds.): EUROBOT 2010, CCIS 156, pp. 149–167, 2011.

2003 at the technology innovation centre (**tic**), Birmingham City University. This was a modular mouse design, aimed at allowing multiple options for hardware and software customization. The modular approach was taken to allow a phased introduction of hardware elements as they were developed. It also ensured the maximum coverage of electronics and computing modules on the existing degree courses, giving the opportunity to assess the ideal requirements. The final version had 5 microprocessors, the master processor, a PIC6F877, communicating by SPI bus to three detachable intelligent wall sensors. These sensors used a PIC16F627 with software SPI. A PIC16C505 processor was used as a port-expander to control an array of LEDs.

ROBOtic had three top-of-the-wall 'wing' sensors, DC motors with simple chopper drive and relay direction change, worm and pinion gear-train, wired and optional wireless RS232 communications, optical position encoders for each wheel, and consisted of as many as seven PCBs. An unavoidable outcome of a modular, expandable design is increased mechanical complexity, high cost, and increased weight. Despite its success as a teaching platform, Robotic was expensive to build, time-consuming to maintain, and limited in performance.

A new design was introduced for final year project students during 2006, called Heretic (Fig.3, Fig.4). Heretic is far more integrated, and designed as a competitive micromouse platform, capable of competing at all events in the micromouse competitions. Heretic has only two PCBs, acting as structural and electrical elements, giving a compact design. The base PCB holds the motor assemblies with spur and pinion gearing, high-resolution encoders for wheel position measurement, H-Bridge motor controller, batteries, charging system, and expansion ports for I2C/SPI sensors. The top PCB holds the PIC18F4520 controller, and has a fully symmetric array of 8 IR-sensors for wall sensing with 4 sensors at the front and 4 sensors at the rear of the mouse. It also supports an RF module and a wire link for RS232 communication with a host PC. It was designed not only to be used in a maze, but also in an open-environment, hence the all-round sensors and the SPI expansion ports for long range IR or ultrasonic sensor modules. The reduction in physical size, and the elimination of the peripheral processors led to a high degree of multiplexing to support the required functionality on a 40-pin processor, however Heretic is a successful design, and is suited to advanced projects and research. One of the Heretic mice was donated to University Polytechnic of Madrid (UPM), as a result of the work of a UPM student placed at BCU on the Erasmus student exchange programme[4].

The evolution of the teaching mice continued during 2009 with project work undertaken by two more UPM students, leading to two prototypes of a simplified version of Heretic. This mouse still supported 8 sensors, but with 6 at the front and 2 at the rear. This version also supported additional sensors by SPI bus, potentially for ultrasonic, long-range IR and also a line-follower sensor. The RF link was removed to reduce cost. One of these mice competed successfully as UPMouse[5] in the Time Trials event at UK micromouse 2009.

---

[4] Raul Lopez Pando, University Polytechnic of Madrid.
[5] UPMouse, Blanca Maria Hernando De Antonio and Christina Martin Ocaña, University Polytechnic of Madrid.

On completion of testing of SMB, sufficient knowledge had been gained to design the production version. This version replaced the RF link with 32K of on-board high-speed data-acquisition FRAM, and 32K EEPROM such that data could be acquired at high-speed from the mouse under normal operating conditions. The rear sensors and SPI expansion ports were removed as they were thought unnecessary for operation in a maze. The shape and size of the PCB was modified to improve sensor clearances, and to reduce overall length, and a number of components were rearranged for a more ergonomic layout. A production run of 25 units was undertaken throughout the summer of 2009, with the majority of the build and test, including writing of test software, undertaken by a summer-placement student[6]. The recent formation of the Faculty of Technology, Engineering and the Environment (TEE) led to this latest version becoming MyTEEmouse.

MyTEEmouse uses 6 independent, narrow-beam IR wall sensors, low-inertia motors with H-bridge control, high-resolution optical encoders, SPI EEPROM and FRAM for data acquisition in the maze, wire-link RS232 for debug output or for transfer of data to/from FRAM and EEPROM, on-board battery charger for the 8 NiMH AAA cells, two push-buttons for user input, two driven LEDs for basic user output, an In-Circuit Debug port, reset button and on-off switch. MyTEEmouse is a robust, low-cost yet high-quality mobile embedded target system that we now use for teaching of Embedded Systems. It is an ideal teaching tool for embedded systems software design, control systems design, actuator and sensor interfacing, data capture and analysis, and data communications. It is fully capable of competing in both micromouse wall-follower and maze-solver categories in the UK competition hosted annually at TEE.

Fig. 1. Robotic

---

[6] James Wilcox, University of Bristol, UK.

Fig. 2. Robotic components

Fig. 3. Heretic concept model

Figs. 1, 2 above show the Robotic platform.

Fig. 3 above shows the concept diagram for the Heretic/SMB platform.

Fig. 4 below is Heretic.

Fig. 5 is SMB – the prototype of MyTEEmouse, which uses the same mechanical design as Heretic, but with significant differences in the electronic design.

Fig. 4. Heretic prototype

Fig. 5. SMB (Student Mouse B)

2   MyTEEmouse Design

Heretic and MyTEEmouse were modeled initially using a 3D CAD package. Motors, encoder and wheel assemblies, batteries, sensors and PCBs were created as parts or assemblies of parts and mated together to form an accurate 3D model of the intended design. Conflicts between mechanical could be easily identified using this approach. The design was exported as a DXF file and imported into the PCB design package, helping to ensure that conflicts between mechanical and electrical/electronic components were minimized.

Fig. 6. MyTEEmouse: Concept

Fig. 7. MyTEEmouse: Final

MyTEEmouse consists of two printed circuit boards – the upper board carries the microcontroller, wall sensors, push-buttons, LEDs, RS232 interface and de-bug/programming port. The lower board carries the motor assemblies, motor drivers, optical encoders, battery charging circuitry, batteries and on/off switch. The PCBs are linked by a 14-way IDC cable.

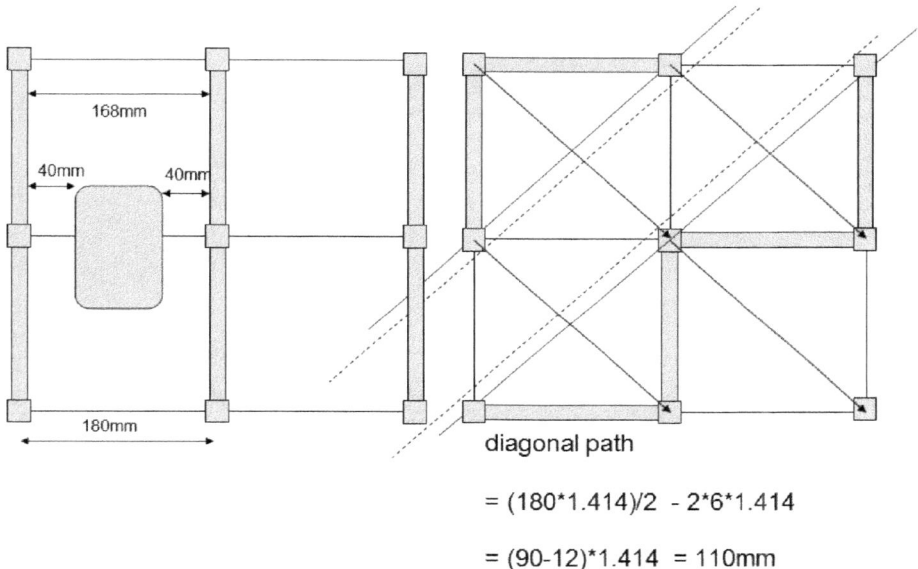

diagonal path

= (180*1.414)/2 - 2*6*1.414

= (90-12)*1.414 = 110mm

Fig. 8. Maze dimensions

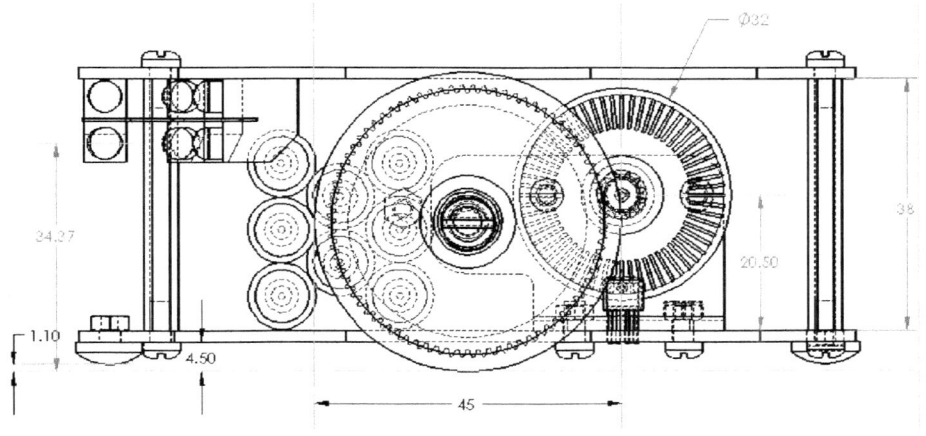

Fig. 9. MyTEEmouse Side View

Dimensions

A micromouse is constrained in size by the maze dimensions. MyTEEmouse is 114mm long and 88mm wide at the wheels. As the maze is 168mm between walls, this leaves 40mm clearance either side when the mouse is parallel to and central between the walls. For diagonals the width of the run reduces to 110mm, giving My-TEEmouse 11mm clearance either side for a straight-line diagonal run.

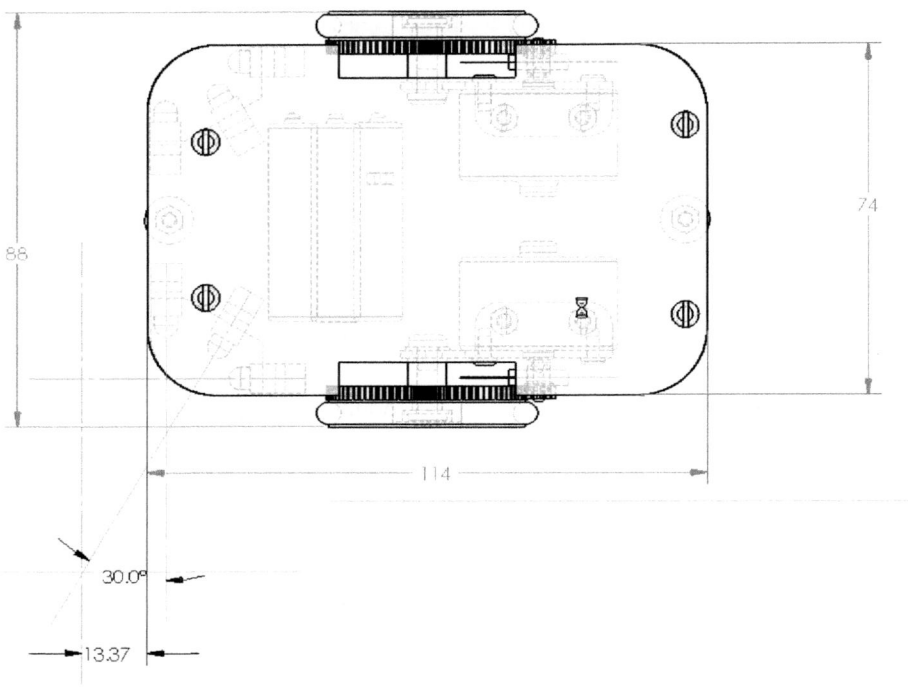

Fig. 10. MyTEEmouse Top View

The drive-train assembly

The drive-train is typically the most expensive element of a mouse design, but for this design the CNC facilities at BCU were used to keep costs low. The motor is a Mabu-chi RF500-TB-12560, commonly used in CD/DVD players. It is a 1.5-12V low-inertia motor (6V nominal) with a relatively low stall current of 350ma and 60g/cm of torque (see technical specification for motor). The motor is coupled to the wheel through a spur and pinion gear arrangement (DELRIN gears from RS components), with a ratio of 80:12. The high gearing ratio gives increased torque at the wheels, and reduces the rotational speed of the motor to give a maximum speed of approx. 1.5m/s in the maze. The 12-tooth pinion gear is fitted with a 60-slot optical encoder disc that rotates between an IR emitter and a times-two quadrature photo-detector (HLC2705) giving 2*60*80/12 = 800 pulses per revolution of the wheel. The wheel diameter is

45mm, hence the positional resolution is 45*pi/800 = 0.18 mm per pulse of the encoder. The bracket that holds the motor and stub axle is machined from aluminium angle. The encoder disc assembly can be simply manufactured using laser-printed acetate discs, two 4mm nylon washers, and the 12 tooth pinion gear. A more robust encoder disc was produced for the final version by chemical milling of 0.254mm stainless steel. Left and right-handed versions of the brackets were CNC-milled from aluminium angle. Stub axles were turned from mild steel bar. Wheels were turned from aluminium bar, and tyres are 40mm internal diameter O-rings. The motor assembly is mounted on the base PCB with two M3 screws.

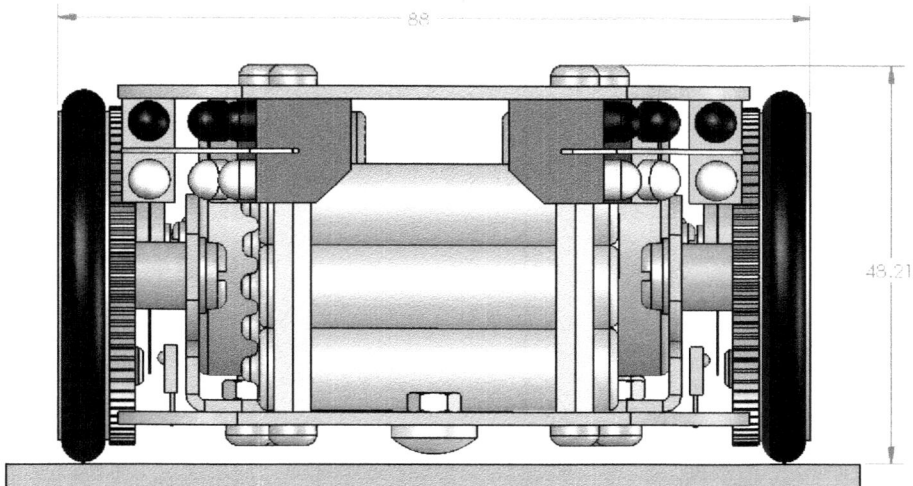

Fig. 11. MyTEEmouse Front View

Multifunctional I/O

The use of independent pulse outputs for the IR LEDs and individual analogue inputs for the IR sensors leaves little I/O for the remaining functions. This was overcome in part by using the internal oscillator rather than an external resonator or crystal (which frees two pins for digital I/O), and also by giving three I/O pins dual functionality.

For example, in Fig. 13 below VS_CS2 allows the user to monitor the battery voltage when it is configured in analogue input mode, or it is the chip-select signal for the FRAM when it is configured in output mode.

The push-buttons are usable when the LED_PBx pins are digital inputs, and in this mode the LEDs are also turned on when the buttons are pressed. When the LED_PBx pins are digital outputs, the LEDs can be turned on and off under software control. This approach offers interesting exercises for lab work, giving a good example of the compromises often required in embedded design.

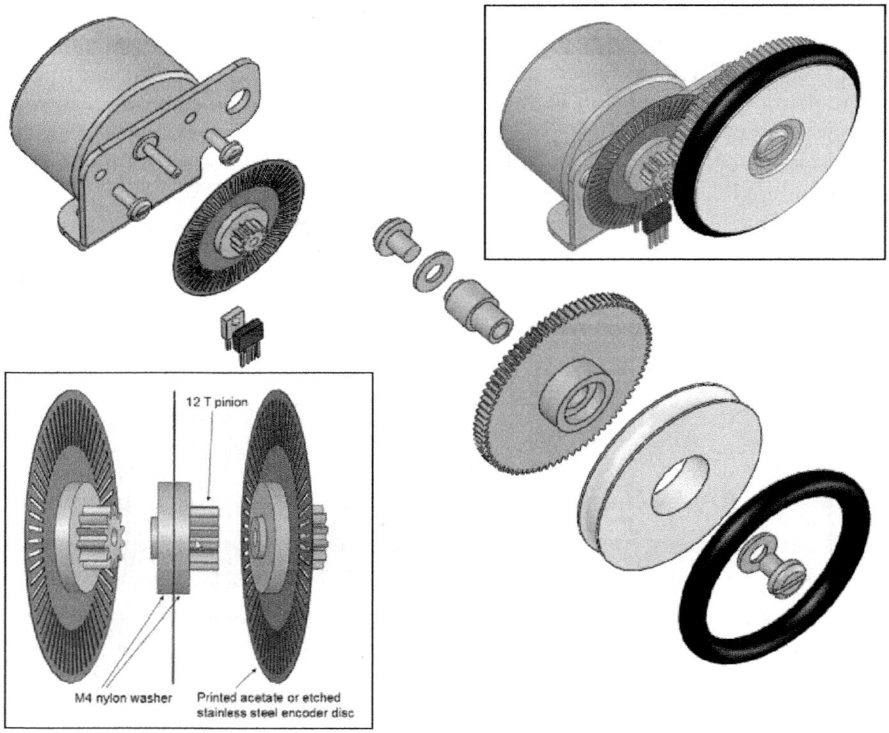

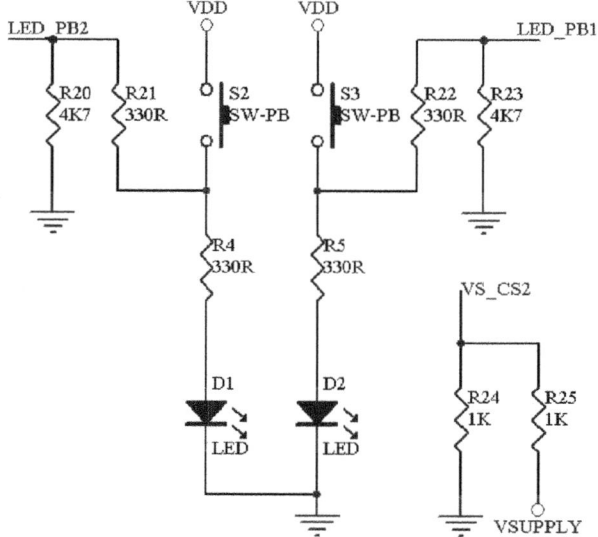

Fig. 12. Drive train assembly

Fig. 13. Multifunction I/O

Motor Driver

The motor driver used for MyTEEmouse is the L293D (rated at 600mA per driver), which is used as a dual H-Bridge. This device therefore supports bi-directional drive of two DC motors with PWM speed control. The PWM signal for each motor is applied to the enable input of the respective H-bridge, while the IN1, IN2 driver inputs control the direction of the motor. Heretic supports both locked anti-phase and sign-magnitude PWM, whereas SMB and MyTEEmouse support sign-magnitude PWM only. This was a design decision to reduce the component count.

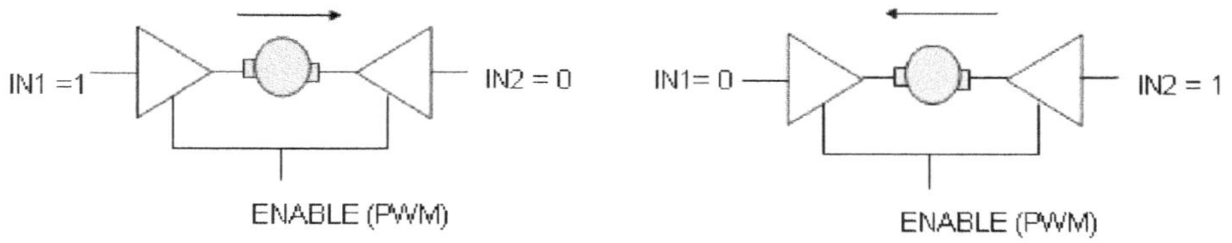

Fig. 14. H-Bridge motor driver

Power supply

The primary power source for MyTEEmouse is 8-AAA 1000mAh NiMH cells, connected in series. The base PCB carries a constant-current/constant-voltage charger which is active when an external supply of 15V is connected and the mouse power switch is OFF. The external supply is also regulated to power the mouse with an appropriate equivalent to the battery voltage when the mouse power switch is ON. This ensures that the mouse will perform similarly when powered from the batteries or from the external supply. A low drop-out voltage regulator reduces the battery voltage to 5V for the logic supply. The motor driver is supplied with the unregulated battery voltage.

IR sensors

MyTEEmouse has 6 wall sensors; two at $0°$, two at $90°$, and two at $60°$ to the longitudinal axis of the mouse. The $0°$ sensors measure the distance to the front wall, and also therefore the presence of a front wall. Typically, the $60°$ sensors are used to maintain a specified distance to the side walls (tracking), and the $90°$ sensors are for detection of wall openings for re-calibration purposes.

Each sensor is formed of an IR phototransistor and an IR emitter, inserted in a low-cost dual LED holder. The phototransistor is used to convert the IR light reflected from a wall of the maze into a voltage that represents the distance to the wall.

The devices used are SFH4503 infra-red emitters with SFH313FA infra-red phototransistors. The emitters are narrow angle devices, while the phototransistors are wide angle, with visible light filtering to reduce the effect of ambient lighting.

Blocking plates are inserted between the emitter and the phototransistor to eliminate crosstalk that would otherwise flood the phototransistor. A specific IR emitter is turned on and the amplitude of the reflected IR is measured using the appropriate ADC channel input.

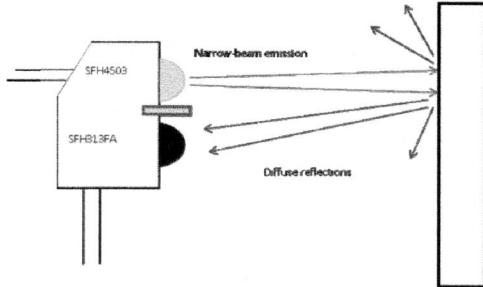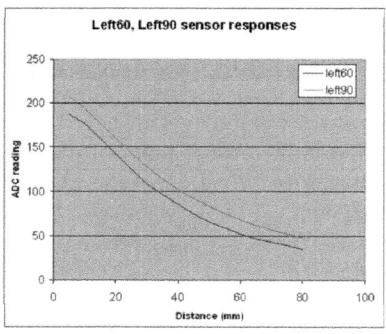

Fig. 15. Distance measurement

Each sensor requires a dedicated digital output port to drive the emitter through a Darlington driver stage, and a dedicated analogue port to measure the voltage representing the intensity of the reflected light. The intensity of IR light emitted by the SFH4503 IR LED is set to be close to the maximum for the device, such that the sensitivity of the IR receiver can be low. This reduces the effect of ambient lighting and/or disturbances. It is also common to pulse the IR LED at very high current – as much as 10 times the continuous current rating - for a short period of time, to further reduce the effect of ambient lighting. This design does not do so, as the design is for student use and inadvertently leaving the IR LED switched on at very high current would cause damage (e.g. under single step/debug conditions).

One technique often used to further reduce the effect of varying ambient lighting is to connect the sensor output to the ADC input through a capacitor. This allows the high-frequency signal pulse to pass, but low-frequency ambient IR is eliminated. The sensors of MyTEEmouse are connected directly to the ADC inputs and an alternative method of 'light-dark cancellation' is used, whereby two measurements are taken, one with the emitter off (dark) and one with it on (light). The dark reading is then subtracted from the light reading in software, giving the required ambient light cancellation. This has the advantage of reducing component count and the readings are less sensitive to sampling-time variations, but it requires two readings instead of one.

Fig. 16. MyTEEmouse sensors and blocking plate

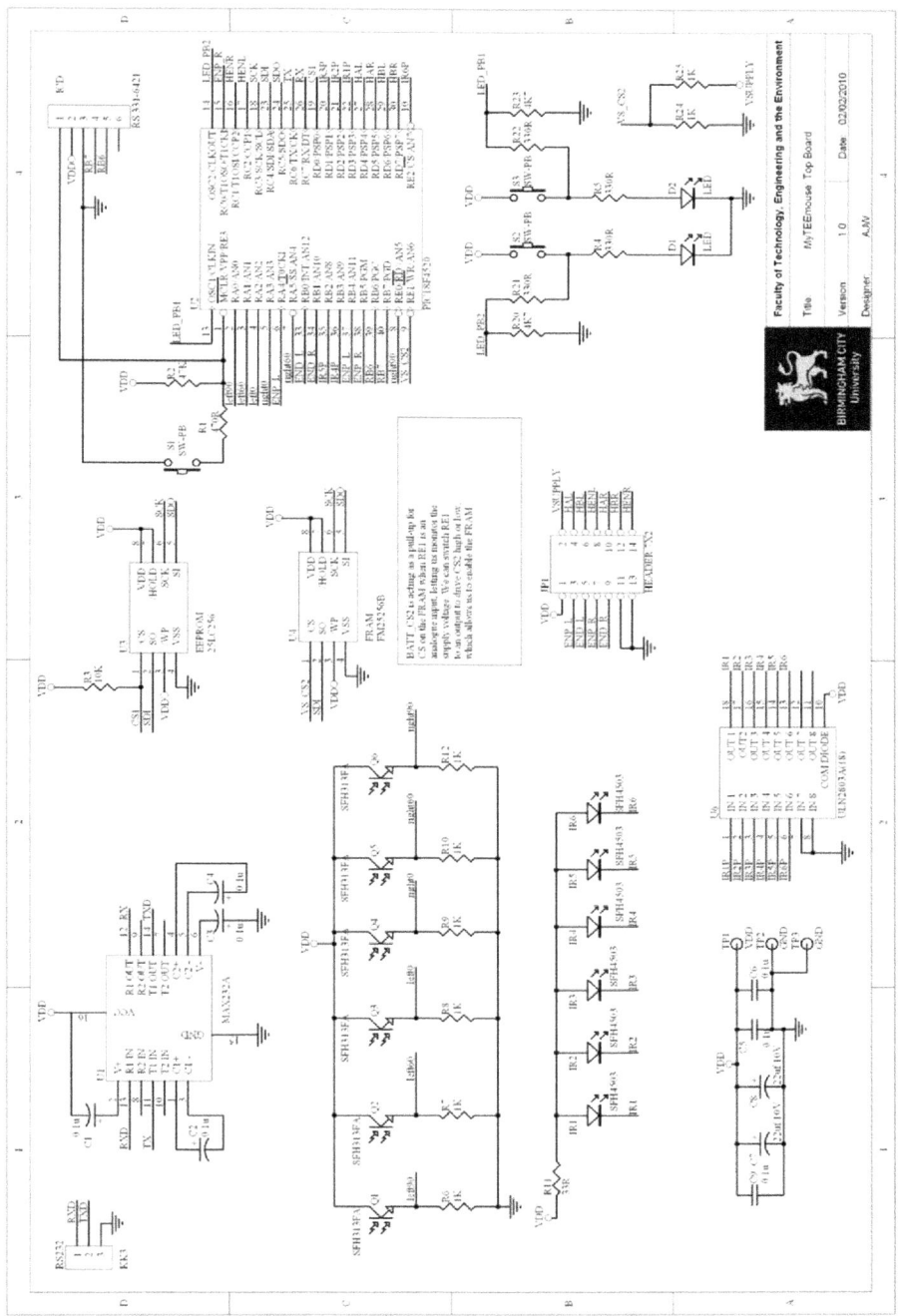

Fig. 17. MyTEEmouse top board

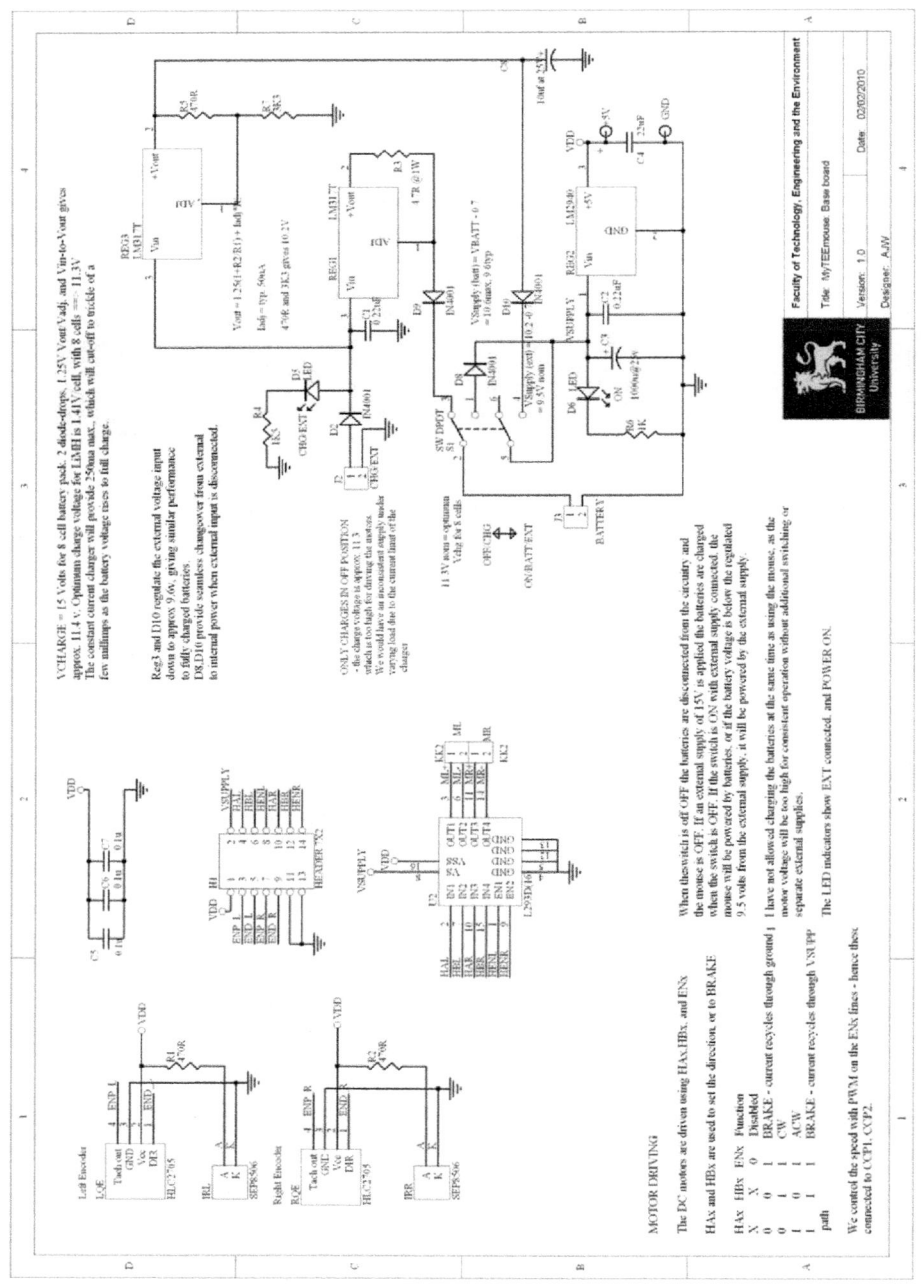

Fig. 18. MyTEEmouse bottom board

Microcontroller

The microcontroller used is the PIC18F4520 from Microchip. This is a powerful 8-bit microcontroller, with adequate peripherals and memory for a maze solver. As My-TEEmouse is an educational tool, the external hardware was mapped to the peripherals to give significant flexibility in the techniques that could be used. The encoder inputs each connect to an interrupt-on-change input and to a hardware timer input. The motor control outputs can be driven using the hardware PWM module, the Compare module, or by directly accessing the port pins in an interrupt service routine. The PIC18F range has a prioritized two-level interrupt system, or it can be used in legacy mode with a single level interrupt. The MSSP module is used in SPI mode to interface to 32K of FRAM for high-speed data acquisition, and 32K of EEPROM for data acquisition, calibration tables, and other storage needs.

Tool Chain

The tool chain used is provided free-of-charge for academic use by Microchip. The MPLAB IDE provides editor, simulator, programmer and ICD debugger interface. The Microchip C18 compiler for 18F devices has peripheral libraries that simplify the configuration and use of the 18F4520 peripherals. In-circuit debugging of C code at the source level is possible with one of the low-cost Microchip or third-party ICD tools such as the PICKit2, ICD2 or their successors.

3 Teaching Embedded Systems with MyTEEmouse

MyTEEmouse is currently used to teach on undergraduate and postgraduate modules. Students are required to undertake a series of lab exercises that lead to a specific endpoint, which as a minimum is the design of the real-time control software for a time-trials micromouse[7]. These exercises draw on the more generic lecture and tutorial material that make up the modules as a whole. A significant amount of control theory is taught in the modules, and a strong emphasis is placed on data acquisition and analysis. Where appropriate, the students are given an introductory course in C programming for embedded systems using static target boards such that the tool-chain is explored and understood, and in particular the advantages of simulation for code debugging (using MPSIM) has been emphasized. The students are then introduced to MyTEEmouse and the following exercises over the course of the modules:

- RS232 communications for data transfer and control.
- Data acquisition to SPI EEPROM and FRAM on MyTEEmouse.
- Pulse waveform generation and PWM for driving DC motors under open-loop control.
- Encoder pulse detection and counting.
- Proportional servo control of DC motors using encoder feedback.
- PID servo control of DC motors using encoder feedback.
- Speed Profiling with an open-loop profiler feeding position changes to the positional controller.

---

[7] Time-Trials rules www.tic.ac.uk/micromouse

- Sensors for distance measurement.
- Wall tracking using wall-sensor feedback.
- Manoeuvres; fixed length straights, 90/180 degree in-place and smooth turns.
- Time-trials: endpoint 1. A time-trials micromouse must navigate the outer-most rectangle of a micromouse maze in the fastest time. This is the normal endpoint for the taught degree module. Students compete in a Time-Trials event to demonstrate the capabilities of their design, and are judged on eleg-ance, control and speed.
- Wall-follower[8]: endpoint 2. A wall follower micromouse requires left, right and 180 degree turns, plus a slightly more complex algorithm than a time-trials mouse, in order to follow a left-hand wall to the centre of the maze.
- Maze-solver[9]: endpoint3. A maze-solver micromouse cannot rely on follow-ing the left-hand wall of the maze – the maze may be designed so that there is no left-hand route to the centre. Instead, the micromouse must explore and map the maze, solve it, and then re-run the solved route to the centre. This is the normal endpoint for the taught MSc module.

Software design

The different endpoints or levels for micromouse robots – time-trials, wall-follower, maze-solver – require differing levels of complexity with regard to the software de-sign. However, while it is relatively simple to design a time-trials micromouse, it is less simple to design so that the software can be developed to a wall-follower or a maze-solver. An example would be that there is no need for odometry to successfully program a time-trials micromouse or a non-contact wall-follower - wall sensor read-ings alone are sufficient to enable the mouse to negotiate the maze. However, odome-try is essential for a maze solver, as it has to know where it is in the maze in order to map and solve it.

Also, the low-level software design for a DC motor micromouse could be expected to be significantly different to that for a stepper motor micromouse, as the nature of the low-level controller is inherently different in each case. However, it is feasible to consider that the high level algorithms for mapping and solving the maze, and the higher level control algorithms, could be identical for even these very different hard-ware platforms[10].

In order to use micromouse for teaching it is useful to have a structured model that represents the different levels of software that must be developed for a maze-solver, and that partitions the functionality sufficiently to identify elements that can be re-combined to deliver the required functionality at lower levels. This may not lead to the simplest solution, but it does give a stepwise approach. This is particularly useful for group projects as the partitioning of the project has already been done, and each element can be built and tested almost independently of the rest. A dataflow model is proposed to meet this requirement.

---

[8] Wall-follower rules www.tic.ac.uk/micromouse

[9] Maze-Solver rules www.tic.ac.uk/micromouse

[10] Informal discussion with Peter Harrison, following Minos '07.

4 The 5-Layer Model for Micromouse Software Design

The model consists of 5 layers:

- **Storage**: maps, tables.
- **Analytical**: high level algorithms for solving, routing, applying heuristics
- **Higher Control**: mouse movement in the maze, speed/acceleration profiling
- **Lower Control**: motor control, e.g. Forward-Rotation PD controller.
- **Drivers**: PWM generation, position counters, distance measurement.

The model is depicted graphically in Fig.19. The Driver layer consists of the lowest level interface functions for wall sensors and motor drivers. These will differ for different motor types, e.g. Table 1.

Table 1.

DC motor micromouse	Stepper motor micromouse
Wall sensors: pulse generators,	Wall sensors
PWM generators (fixed frequency, variable duty)	Pulse generators (variable frequency, fixed duty)
Position encoders	Battery monitor
Battery monitor	

The Lower Control layer for DC motors would consist of the PD controller(s), or other controller implementations, that would give control of speed and position of the motors. Stepper motors would not require much at this level – they are inherently position controlled, effectively combining the functions of the Driver and Lower Control layers.

The Higher Control Layer is where the movement of the mouse in the maze is controlled. There may be more than one controller at this level, but not necessarily active at the same time.

- The profiler applies a specified acceleration/speed profile to the motor, stopping at a specified position.
- The sequencer would replay a series of profiles from storage (e.g. a move array).
- The mapper is the controller that takes sensor readings, puts the information into the maze map, calls the solver, then applies an appropriate profile based on what the solver returns.

The mapper and sequencer would not be active together – and could have been considered to be part of the solver. However, partitioning the functionality in this way gives simple, easily testable controllers. For a maze solver, the sequencer may only be used for testing the profiler, for example reading moves from a dummy route array to test the accuracy of the positional control, or the stability of the motor controllers. For a time-trials mouse (which has a known route) the sequencer could become the major controller.

Micromouse Maze Solver
5-Layer Data-Flow Model

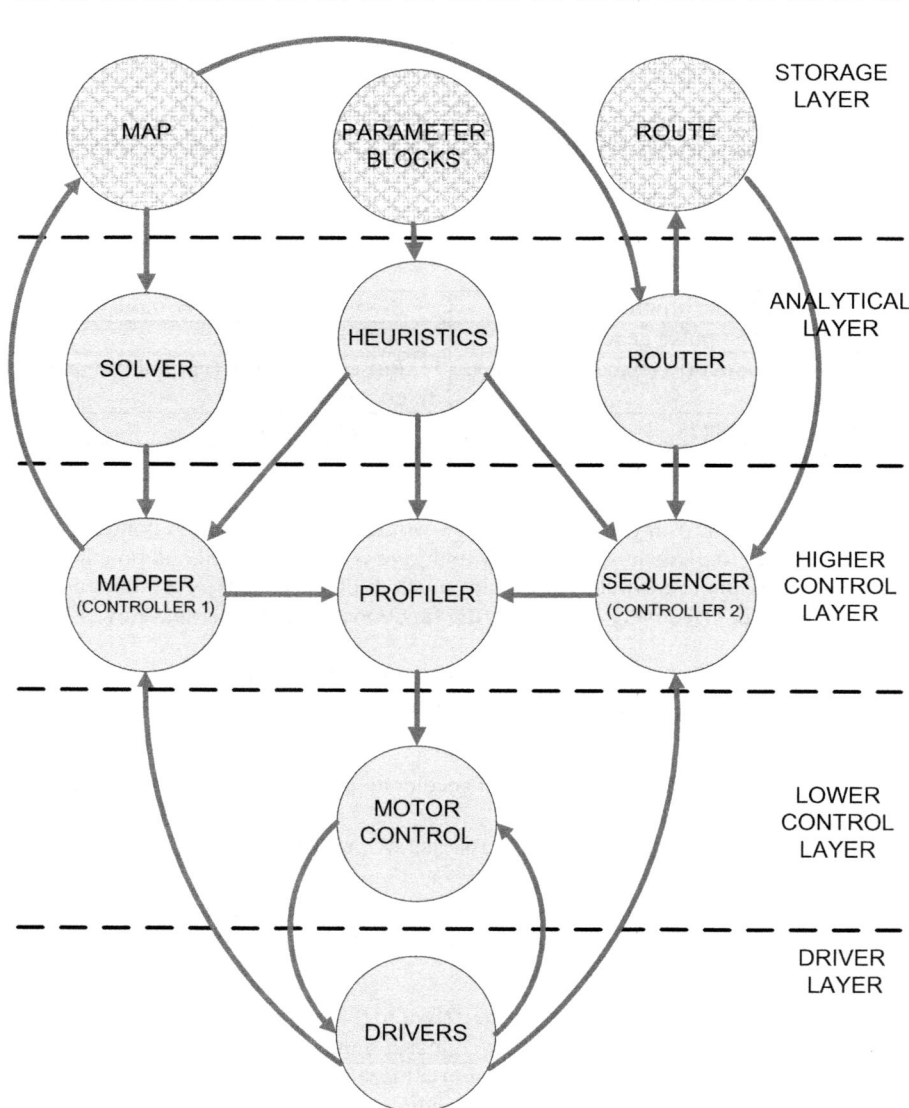

Fig. 19. The 5-Layer Model

The Analytical Layer consists of the solver, the heuristics, and the router. The solver and router should now be independent of the hardware platform, with the heuristics doing the necessary matching. The solver is an algorithm that identifies the best move to make next. The router's job is to determine the optimum path once the solver has completed its task of mapping the maze – possibly choosing from several paths to the centre. The heuristics feed both solver and router, amongst other things such as switching from in-place to smooth turns, and increasing speed and acceleration on successive runs, drawing the new parameters for the profiler from the parameter blocks. Again, partitioning the analytical tools in this way simplifies the task of designing code and debugging it.

At the simplest level of implementation, much of the structure of this diagram falls away. A simple time-trials mouse following a known track requires only sequencer, profiler, motor control and drivers. A wall-follower would require a simplified mapper, profiler, motor control and drivers. The profiler at its simplest would not require acceleration/deceleration – the mouse could run at a slow constant speed, with speed/direction change on the wheels for turns.

The model describes a structure for micromouse software, and provides a stepwise, bottom-up approach for building micromouse applications.

Summary

MyTEEmouse has been in use for 6 months. A total of 25 units were manufactured, and the 20 units released to the lab are in regular use by 40 students. The design has been proven to be robust, but above all the consistency of operation from one mouse to the next is excellent. The 5-layer model has been presented to and been well-received by students, and has been adopted by several project students for implementation of wall followers and maze solvers.

It has been said that the micromouse problem has been mostly solved[11] – however solving the problem for oneself is still an excellent learning experience.

---

[11] Braunl, Thomas, Embedded Robotics – Mobile Robot Design and Applications, Springer, 2008.

Autonomous Mobile Robot Development in a Team, Summarizing Our Approaches

Andrey Demidov[1], Andrey Kuturov[2], Anton Yudin[1], Boris Krasnobryzhiy[1],
Mikhail Chistyakov[1], and Rustam Borovik[1]

[1] Bauman Moscow State Technical University, IU4 Department,
2-nd Baumanskaya st., 5, 105005, Moscow, Russia
[2] Lebedev Physical Institute of Russian Academy of Science,
Division of Solid State Physics
Leninskiy prospekt, 53, 119991, Moscow, Russia
{the-admax,kuturov}@yandex.ru,
{Skycluster,mlazex,borovikrustam}@gmail.com,
capitan_Flint038@mail.ru
http://www.bearobot.org

Abstract. This article summarizes experiences of beArobot team members while developing mobile autonomous robots for Eurobot competitions. This text is an attempt to systemize approaches needed to successfully finish year-long project by a small team. Text covers method of formulating design tasks based on early robot strategy development from the plain text of competition rules. Further analysis of mechanical parts commonly used from competition to competition and unique mechanisms cover "how to make" and "where to get" questions. Basic principles of mounting electronic parts on mobile chassis and unified software platform approach for rapid modular program development finish the list of topics covered by the article.

Keywords: Eurobot, mobile robot, design, project management, good practices.

1 Introduction

21st century is no longer the time of individuals, it's the time of small teams of 4 to 7 members [1] able to solve any problems world is demanding including robot production and research. But how this is different from what was before, the impatient could ask? And the answer is in our educational system. Authors experienced engineering and fundamental university education but none included questions of team work.

You can never understand something until you have experienced it. This is true for humans and this is true for knowing what team actually is. Interested to get experience? One of the best solutions for engineers for now is to take part in team competitions like Eurobot [2]. Experience is guaranteed but what about its quality?

All teams are different like people but there are always traits that could be called basic. These traits allow some teams to win and others to lose. And there is nothing good or bad about that – as it has already been said we are only on the threshold of the

D. Obdržálek and A. Gottscheber (Eds.): EUROBOT 2010, CCIS 156, pp. 168–179, 2011.
© Springer-Verlag Berlin Heidelberg 2011

age of new engineering education. Hopefully with time we'll see courses on team work included in university schedules. But for now we should accentuate two main elements that will be in the center of those courses – people and cooperation. These two define success of any project. People bring knowledge and this part is pretty well covered with present education. As for cooperation there is still a lot to be done but a positive trend is that this process can become a skill and there are plenty of such examples if you take football or any other team sport.

Bearobot is a team with 2 year Eurobot history already with all traits of a 2-year old child. No exact knowledge or winning strategies but some experience and wish to share. We hope this article will allow us to define first steps to methodology of team work and team cooperation basics first of all for ourselves and for those who's interested.

2 Start of Project

Being a contest of teams presenting their mobile robotic solutions in a sporting spirit Eurobot is more than just colorful matches between two autonomous robots. It's a bridge to new level of education. Education including team work skills along with outstanding potential for practical knowledge and research. Competition on one hand and cooperation on the other both stimulate perfection.

When speaking of Eurobot project several parameters should be defined to understand what it is for participating teams. First parameter is time - project is limited in terms of time and usually it is given about 6 month to finish it. Second parameter is the number of team members. There could be different points of view on optimum number of people but our team believes that this number should also be limited at its maximum of 7 members. Actual number of people is changing with years but what proved to show tendency is the number of core forming participants - three is enough to finish the project. And another 1 to 4 are usually new comers.

When project is at its starting phase and team is gathered there should be a general plan template ready to organize its work. Such a plan reflects Eurobot specifics and evolutionary progress of the team while core members gain more and more experience and abilities to teach new comers and form engineering culture of development. Let's revise its main outlines.

For Russian students academic year starts in September. Teams usually would have about a month to search for new members until rules of Eurobot are published by early October. After that, project actually starts.

First step to make is obligatory elements and field construction for new members to feel the team and to start with simple and yet demanding tasks. Concurrent process includes brainwork on main robot strategy. Usually there will be 3 different tasks to be engineered with new mechanical solutions. On this stage strategy includes mainly analysis and estimation of task complexity. In future when it is needed to decide which task to actually realize an instrument of comparison is needed. This time we have to remember that competition goal is victory, and victory is based on points gained by the team. So naturally points will be the first parameter to analyze. But is it enough to form a strategy? Points usually would reflect complexity of the task determined somehow by the organizers. On the other hand one needs an independent parameter which would reflect real complexity specifically for his team. Such a parameter could be time to collect single playing element to score. When transferred to these 2 parameters all the tasks can be formally operated and thus we can define more suitable tasks for the team in an objective way. Let's call those 2 parameters a solver system.

To clarify first step estimations based mainly on member experience the actual development of mechanisms should be done. This step could be called prototyping as one has to "invent" a solution to a task and then test whether it works or not. When done properly this step brings accuracy to the solver system adjusting previous estimations.

Next step is actually another iteration of strategy forming based on accurate solver system. This time final decisions are made on which tasks to solve and which prototype to include in robot construction.

After all preparatory steps actual work starts. We suggest using distributed control system in the project. It allows true concurrent development of system parts by team members and forming reusable modules for later applications.

In case of modular design integration step is very important. Sometimes physical rework of parts may be needed but mainly this step forms true basis for final control realization and configuration. Ability of the team members to accurately and professionally organize robot control and module interaction determines success of the strategy and ideas put into system parts. Conventional wisdom even states that all possible strategies one could think of can bring victory to the team. Only strategy realization makes this statement come true or false in real life. That's why in the next section we'll discuss practical questions of robot building.

3 Growing-Points

This section is called growing-points mainly because it describes all that knowledge gained by teams while progressing on Eurobot projects. Actually there can be defined 3 such points: mechanics, electronics and programming. Of course each point is bottomless in details and can be divided into less powerful aggregations with other names. But we are going to concentrate on general trends forming distinctive abilities and skills in competition participants.

This section provides information and personal beArobot team experience based on Eurobot 2010 rules [2]. For better understanding of further subsections it is advised to consult the rules. We believe this section to form basis for future research in our team and hope to attract interested individuals and teams to discussion on presented questions.

3.1 Mechanics

Generally from mechanics perspective robot for Eurobot competitions should combine common movement functions and a number of additional acting mechanisms. Eurobot rules change each year and this leads to changes mostly in acting or in other words manipulating schemes. Chassis and navigation system once correctly implemented do not need any significant changes. This fact brings modularity from the very start of another project for a team. For mechanics development it should be common to whenever possible to produce parts able to be transferred to other robots of the team as well.

Robot easy to manufacture is rather difficult to design. Ideal robot should consist only of standard parts, available within different production sectors like automobile, furniture, modelling sports, etc. Such a robot will possess a number of advantages:

- Easy production – accurate and detailed design of mechanism part is not needed, as commercially available part is usually optimal and reliable in design;
- Reliability – when constructed with reliable parts, quality of which is proven by years of production;
- Simple assemblage – mainly because of completeness of parts being used, freeing from part's mechanics workout;
- Repeatability of assembly – using unified parts makes it easy to get to modular design, which will allow easy repeating of modules.

Practically it is almost impossible to design ideal robot. Often robot would only partially consist of standard elements. If we look at the means of realization of any mechanism for Eurobot competitions we could define the following list of possible options in order of decreasing preference:

- Usage of commercially available parts;
- Self-design parts and order professional production;
- Self-design parts and further self-realization.

The quality of resulting robotic solution is fully dependant on experience, skill and knowledge of mechanic in a team. Not to start all over again with each new project it is good to track successful solutions whether they are a product of a team itself or whether they are a product of team's interaction with other teams on competitions and other activities. Understanding the impact of mechanical solutions on the whole robotic project, analysis of mechanisms commonly used on Eurobot robots seems interesting, challenging and useful idea we hope to realize in our future work.

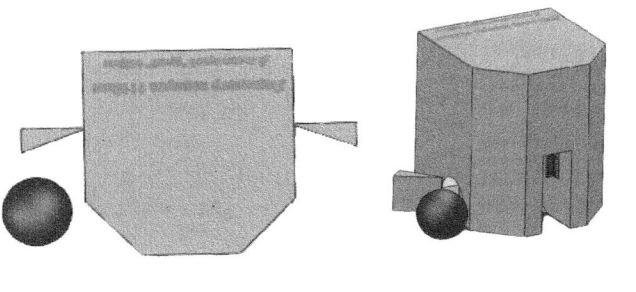

Fig. 1. Robot for Eurobot 2010. General view

Let's analyze fundamental principles for constructing robot mechanics by the example of Eurobot 2010 rules. It is to be said that from the very start we decided to concentrate only on corn collection. We believe it is not necessary to review and analyze each and every part on the robot but for a single mechanism of collecting and storing corn inside the robot in this case.

For general concept of the robot let's refer to figure 1. Left side of it demonstrates robot from top view, and right side – axonometric projection of robot body. In the figure you can also see several playing elements included for size reference purposes.

Fugure 2 complements more information on how robot systems are designed and reveal general ideas. Right side of the figure presents more general front view and left explains how inner mechanisms are organized. You can see numbers in the pictures. These numbers are to help locate parts in our further system explanation.

Main idea to manipulate corn elements consists of several parts. First part is under number 4 in the next figure. It's a sort of brush which main purpose is to bring corn down. Fake corn is able to come through robot interior and thus eliminating the need to avoid it while cruising in search of playing elements. Number 1 represents sensor array that allows detection of corn. Whether it is fake or real corn element – control system is able to make distinction with original sensor placement. Number 3 represents grabbing system of the robot. When real corn is detected elevator is activated grabbing corn and transporting it to the top of robot body to release in one of two storage containers. Finally number 2 represents supplementary ability to push balls towards the scoring area.

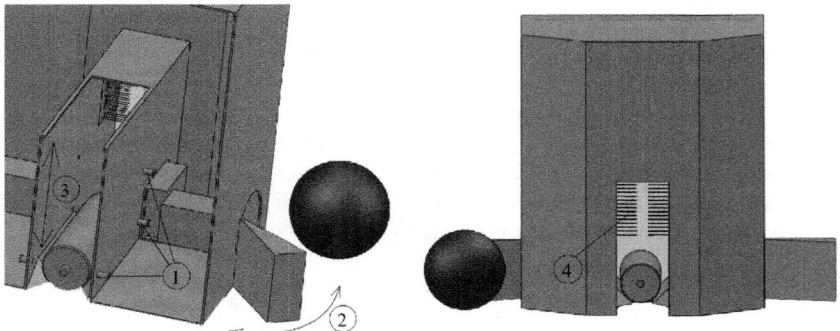

Fig. 2. Robot for Eurobot 2010. Collecting mechanisms basics

Let's review corn collecting mechanism. Its purpose is to grab lying corn from the field and to get it to the storage containers onboard. One of the important tasks of the mechanism is precise positioning of corn as requested by storage and unloading mechanics. Let's define corn for processing as lying in the open-ended passage (which is wider than corn diameter for 15cm).

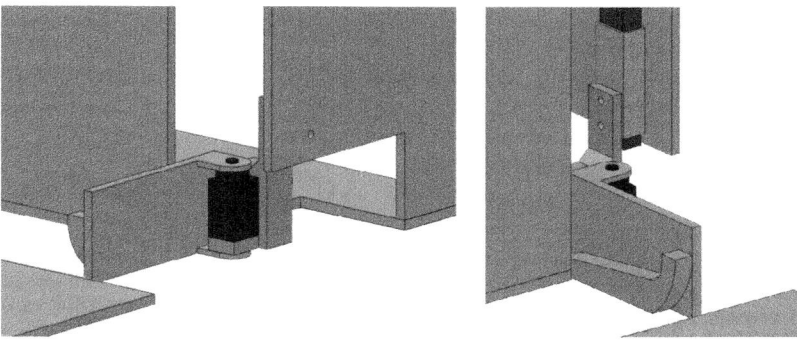

Fig. 3. Corn collecting mechanism. Door for precise corn positioning.

Main elements of the grabbing mechanism are two doors able to open and close with the help of servos. On figure 3 you can see closed state of one of two doors needed for positioning. When corn is detected by sensors doors are brought to closed state. While closing, doors can slightly move corn element thus positioning it between two doors in a known state. Door and servo are fixed to guidance track allowing both to travel vertically. Vertical movement is due to flexible wire system activated by motor. Let's analyze origins of parts in described system.

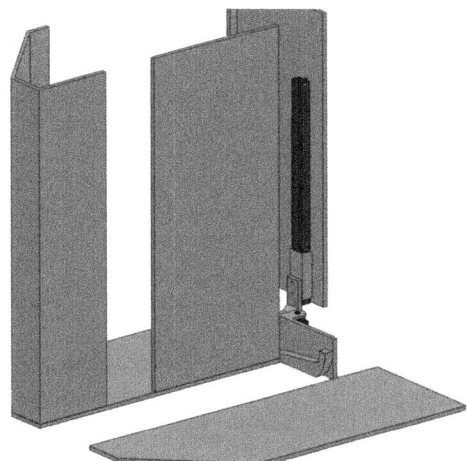

Fig. 4. Corn collecting mechanism. Door for precise positioning.

Servos and motors were bought as they are freely available on the market. Some effort was applied to find suitable track and wire, but they are also bought elements.

As doors are original in design they had to be produced by the team. Main material to process with laser cut machine was PVC. Rollers for wire guidance were produced with 3D printing machine. As one can see while constructing grabbing mechanism we used commercial and self-made parts, produced with modern machines kindly provided by the Russian Eurobot NOC [3].

While system construction we discovered several specific mechanical features to be highlighted. Suitable guidance track was hard to find. A good choice appeared to be a part from writing table's drawer move system. Several guides from different manufacturers were tested and best was chosen for our purposes. Wire for vertical movement of the carriage was chosen from fishing equipment. Special flexible and strong fishing line was used for that purpose.

Rollers satisfying our needs for compact design could not be found and thus decision was made to construct them individually.

Described examples show our suggestion to use commercially available parts, limitation factors being search time and availability of suitable designs. Depending on both factors self-development may be required.

3.2 Electronics

Electronics and electronic parts have a long history. If we look back at those rates of growth in 20th century, according to so called Moore's law [4], which are still true for the whole field, we'll see a never ending trend of evolution of electronics.

Some advantages of such development include simplification of usage of different parts which some time ago were considered too complicated to be used in ordinary design. As a consequence we have a large amount of electronic devices to choose from these days.

If we take electronic design in terms of Eurobot competitions we'll see correlation to previous statement and to the large choice. Naturally each team would have its own thinking of what is best for their current robotic solution and thus one can see a lot of differently organized electronics onboard of robots. We are not going to advance in the question of what is the best and the only good practice to arrange electronics for now, but we are going to discuss our way of doing it and ideas lying beneath.

When you get in touch with Eurobot for the first time you see competitions. From spectator's point of view the reason to come is to see challenging opposition and colorful matches. For him it is natural to expect each team's aim to be victory. But then if we take a look at the same question of participation from competitor's perspective we'll see that it is more of education in place. We'll see Eurobot as an educational project.

Now if we turn back to electronics taking into account educational perspective we would suggest among other variants of buying electronic parts from any third party to design it within Eurobot project as other mechanical or software parts.

Now what about difficulty level? Robotics has immense potential for invention and thus each solution can radically be different from others. Depending on the problem to be solved electronic solutions will be of different difficulty. The good point is that these solutions can be simple and thus designable by the new comer from the very start.

For our team it is usual to use 8bit microcontrollers of AVR-core nature. Having one microcontroller with 32kB of flash is enough to organize simple navigation and sensing needed to avoid obstacles. As a matter of fact our experience shows that that is the minimum system requirement to be able to take part in autonomous robot contest.

To be able to progress with time it is good to save previous solutions and use them in new projects. This way any team can become successful in several years if there is no source to learn from other than bare competition and preparation experience. For that idea to work it would be necessary for electronics to provide a means to independent developing of devices and their easy interconnection in the future. Each device can be constructed by different member of a team or at different times. A good practice in such a case would be sole bus selection made from the very beginning. Our team prefers CAN bus for that purpose.

With time growing experience for team members will bring more interesting ideas to electronic organization of course. Some key aspects of choosing approach to electronics architecture design are: work organization of team members, skills needed to work with the system, maximum abilities and functionality of the system, scalability.

In our case we prefer modular approach to electronic design. This allows making common parts identical and to design or redesign only those parts which cannot provide means needed for current project. Such general electronic part is shown in figure 5.

Fig. 5. Microcontroller embedded board used for control on the robot.

Question that is closely connected to bus or, more generally, to architecture selection but is more practical is the number of wires in the system. It is obvious that the less wires are on the robot the better. Large number of wires means more work while mounting and even more work while debugging if something goes wrong. Sometimes this would lead to different influences of noise effects preventing the system from correct operation. How to make the number of wires small? First measure is to introduce differential buses like CAN physically consisting of a definite number of wires. And second measure is to put separated microcontrollers as close to their field of action as possible connecting them with a bus for communication.

Finally if we touched electronic mounting onboard of mobile robot let's research how to do it right. Usually questions of electrical noise effects on a mobile device are not trivial. Such effects have several origins one of which is appearance of static electricity. As such effects could influence normal operation of the whole system or even damage parts, we should name a number of rules to be fulfilled while designing electronics mounting scheme.

Problems of static electricity arise when designing aircrafts (moving airflows different for parts of the fuselage cause difference in electric potential) or when designing powerful antennas (formed by antenna electric field could make parts of house interior also different in electric potential). Special and yet simple measures to discharge the body of a trolleybus to ground can be easily observed on the road. You can see now that such questions are rather common for a range of devices. Let's investigate more on what causes such problems on mobile robot.

One of the causes of static charge on the robot body experienced when touched by hand could be triboelectric effect [5]. Most popular materials used in Eurobot robot design are aluminum, different plastics, acrylic glass and wood for the body, rubber and polyurethane for wheels, acrylic paints are used for main playing field, and human skin as from time to time contact. All these materials have different electrochemical potential. The row from most positively charged to most negatively charged will be: human skin, +, aluminum, 0, wood, -, acrylic, rubber, -, polyurethane, plastic ("+" or "-" indicate large gap of potential between elements). Materials brought into contact will attain different charge – positive for those closer to the start of the list, and negative for those closer to the end.

Usually when robot is placed on the playing field one can expect acrylic paint to cause electrostatic effect when in contact with polyurethane wheels. This effect is much less strong if wheel material is rubber. While moving on the field robot tends to collect charge if there is no special means introduced to its design.

Let's review what the consequences of such physical behavior are. If this question is left unsolved by the designer it can cause sparks and unpleasant sensation while touching the robot with hand. Another more technical aspect includes ability to cause damage to microelectronic parts on occasional discharge. This leads to the need of some sort of protective package for most sensitive parts and thus to more work. And finally excessive charge forms electric fields able to cause noise effects of different sort in electronic schemes. This can lead to unpredictable and difficult to track and debug unstable functioning.

First solution is to keep charge separated locally in place of contact. This solution is simple and pretty straight forward. Its main idea is closely connected to the physics of triboelectric effect and states that when charges on both surfaces equalize each other there is no more charge gathering in place. In this case we have to put dielectric as close to the contact zone as possible to prevent electrons from travelling along metals used in design.

Second solution is to transfer gathered charge to onboard battery along metallic elements of the robot body. When most of the body is made from metal it is easier to connect main battery "–" electrode to it. This way we let electrons to flow easily to and from energy source. As long as we use polyurethane wheels electrons will be generated on their surface. This will help "-" electrode of the battery and will not waste its energy on compensation of supplied potential. On the other hand if contact zone is presented with metallic roller it will tend to gain positive charge and thus will waste some of the battery's power. Both effects seem to be negligible but need to be researched in future work.

Finally it is to be said that a good practice is to ground the body of a robot, connecting metallic parts together and to the ground of the battery. This measure allows more predictable system operation as noise generated by the robot itself with its electronics and static charge gained while moving is neutralized. More strictly it provides noise immunity to control systems of the robot. Additionally this measure allows neutralization of sparks and harmful discharges between robot and its environment.

3.3 Programming

Currently, the hardware platform of our robot consists of similar modules based on AVR MCUs, namely AtMega128. Each of them serves particular peripheral devices (e.g. sensors or motor drivers) and is connected to common bus used for data exchange. The conditions of work are rather harsh in aspect of various noise and limitations (i.e. time and energy limits). According to this further on we suggest a concept of robot control system.

The logic of robot is built via "atomic" blocks – functions. They form robot's behavior by connections between themselves. Actually robot consists of electronic modules, each of them executes a number of functions, e.g. controls motors, processes signal from sensors or prevents collision with obstacles.

These functions can be randomly connected to each other according to neuron-network construction principles: output signal of a "neuron" can be transmitted to several destinations (one-to-many).

Linkage between functions is organized at system setup and in some cases when already in use. The best configuration for functional blocks satisfies all stated network requirements.

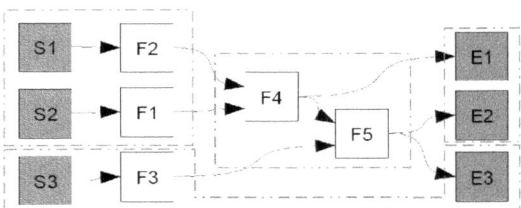

Fig. 6. Possible outline of control system. Sensors (S1-S3), data processors (F1-F5), executors (E1-E3). Single hardware modules are marked with dotted line. Arrows show data flows.

A platform is needed that will make it possible to implement such structure of robot control. Actually it is implemented as shown on figure 7.

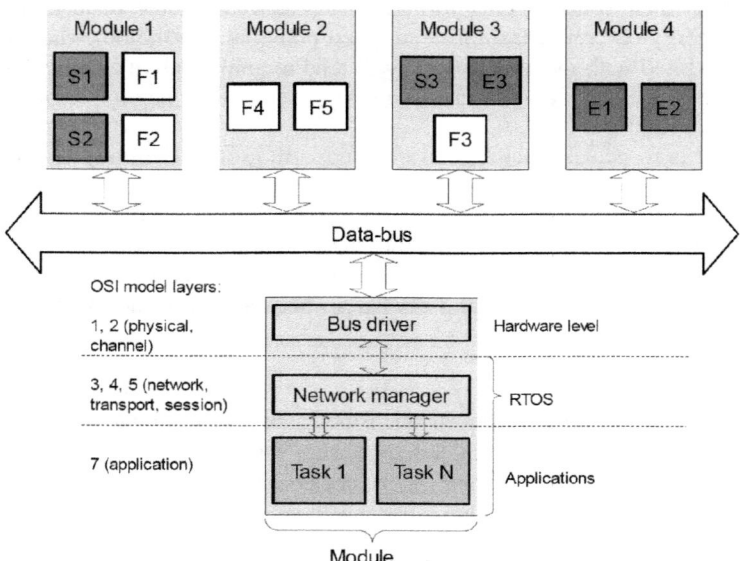

Fig. 7. Control system structure.

It is necessary to separate actual network structure from its logical representation. Implementation means installing RTOS (real-time operating system) with multitasking support on MCUs that provides generic and platform-independent programming interfaces to hardware, network and MCU's peripherals.

Such OS (operating system) should be:

- Flexible (because we want to use the same operating system on every controller despite their functionality destinations);
- Tiny (considering the fact that applications need a lot of resources, especially memory);
- Functional enough to provide any needed routines at different levels of abstraction.

Finding such an OS frees us from writing any trivial code of our own and greatly increases time for essential tasks. Let's take a look at some of such systems, namely Nut/OS [6] and FreeRTOS [7] and choose the best for our purpose.

First of all, both OS support AVR MCU family and its capabilities: timers, interrupts, IO-ports, internal and external memory and other functionality. Also they have API for multitasking (managing tasks, inter-task communication routines, etc.). But Nut/OS as opposed to FreeRTOS has also memory management routines, better bus support (CAN, UART, ISP, TWI...), device support (ADCs and DACs, timers and other peripherals), network protocols (IP, TCP, PPP, ICMP...) and filesystem support (FAT12, FAT16, PHAT). Besides it provides configuration parser that produces makefiles and header files with different options for target OS's kernel and modules.

Within all advantages and power of Nut/OS there is expectable redundancy in functions it provides from our perspective of robot control. In particular we believe file system means of presenting every possible actor (e.g. devices, data, and processes) as a file are not crucial for automatically and autonomously controlled systems, actually this approach introduces a level of abstraction we could ignore in a definite system, but sometime later they could come in handy of course.

For now we will ignore most of excessive functions provided to carry on with research of existing real-time systems, their abilities and appropriateness for robotic tasks. In our opinion both FreeRTOS and Nut/OS are documented and coded well, what makes them a good start. You can even find comprehensive guide to FreeR-TOS's internals on its website that we found useful for studying advanced programming and multitasking basics.

At last we came to conclusion that it's better to try Nut/OS, due to its rich set of drivers and interfaces. Of course if it proves to be suitable enough we still need to introduce a lot of modifications to it to actually serve our purposes. Hopefully we believe to discuss our experience in the next articles.

4   Conclusions

This article summarizes experience and understanding of Eurobot competitions gained by still young beArobot team. All presented information is fruit of labor of many current members of the team but of course the influence of those once in a team will remain within team spirit and culture. With this article we hope to start a series of arrangements to bring more order to Eurobot projects at least for Russian teams.

We slightly outlined future research topics and points for professional growth. Besides well known topics like mechanics, electronics and programming we believe team work skills to be crucial for success of any Eurobot project and Eurobot movement in general. Some practical questions and examples were presented in the article to explain our current project and those methods and approaches we believe to be good practices in Eurobot design.

References

1. Belbin, M.: Management teams: Why They Succeed or Fail, 3rd edn. Butterworth Heinemann, Butterworths (2010)
2. Eurobot, international robotics contest, http://www.eurobot.org
3. Eurobot National Organizing Committee (NOC) of Russia, http://www.eurobot-russia.ru
4. Wikipedia article, http://en.wikipedia.org/wiki/Moore's_law
5. Jonassen, N.: Electrostatics, 2nd edn. The Springer International Series in Engineering and Computer Science, vol. 700 (2002)
6. Nut/OS real-time operating system, http://www.ethernut.de/en/firmware/nutos.html
7. FreeRTOS real-time operating system, http://www.freertos.org

Distributed Control System in Mobile Robot Application: General Approach, Realization and Usage

Andrey Vlasov and Anton Yudin

Bauman Moscow State Technical University, IU4 Department,
2-nd Baumanskaya st., 5, 105005, Moscow, Russia
vlasov@iu4.ru, skycluster@gmail.com
http://www.bearobot.org

Abstract. This article is aimed to give a general idea of how to organize robot control with a distributed network of embedded devices. It is an attempt to formulate simple concepts that will allow easier cooperation for team members during development. Suggested approach when properly described, understood and implemented is expected to form a good platform for concurrent individual software and hardware development and for educating new team members. Giving flexibility such an approach leads to convenient modular robot design. Realization of proposed approach is also introduced with a hardware version of embedded controller board, being the "heart" and the "brain" of any module on the robot.

Keywords: Eurobot, mobile robot, distributed control system, modular systems, team development.

1 Introduction

It's common to hear from different people now and then the same question about new technologies expressing fear or lack of trust in evolutionary scientific process. There's hardly a thing we can tell them to reassure that things are going their natural way but to name the reason we are doing this global research for. And the reason can be found deep inside any of the living creatures' mind. Being an open living system brings a strong urge to put order in the chaos of life. Order can be thought of in terms of construction as an engineering process – the only process to check whether our knowledge and thinking are correct and fully reflect nature of things. In other words we cannot predict anything until we have experienced it, built it.

"Who are we and what are we doing here?" – is also a question that rises every now and then in us. The true nature of such questions can hardly be known at our current state of understanding. But as many other facts which we cannot explain they exist. And the only way to find answers is to search the outer world and to get to know ourselves better step by step. Researching and building robotic systems could be one of such ways as they are similar to living beings in many aspects of existence.

Speaking of artificial systems we could turn to Herbert Simon [1]. In his book he explains how artificial becomes a part of natural. And more precisely he explains and demonstrates that human thinking includes natural and artificial formations. There are

D. Obdržálek and A. Gottscheber (Eds.): EUROBOT 2010, CCIS 156, pp. 180–192, 2011.

no motives not to believe this and many other scientists. And if we take these ideas as true we could reveal many mysteries in ourselves by constructing more and more intelligent machines. At least the artificial part of our own self could be finally understood.

Hopefully ideas briefly expressed above will help to explain the reason of the research on suggested topic. Throughout the history of science many efforts were made to make Nature clearer. The work partly described in this article is a small contribution to this movement. Concentrated mainly on control processes and communication its final destination is to describe a system similar to human nervous control system, highlighting important mechanisms forming intelligent behavior.

It should be noted that this article indicates an early stage of research and thus is also aimed at finding interested opponents and colleagues to discuss and evolve the suggested approach.

2 Why Distributed?

Luckily having participated to several robot projects for Eurobot competitions the author experienced several ways of organizing control on a mobile autonomous system. Each project was different in a center topic of development. Approaches ranged from a complicated control system using a compact PC with its power to calculate and style of higher level programming to a simple embedded microcontroller as an attempt to find more simple ways to organize the whole robotic system. All control systems realized no matter how complex they were contained serial approach to data processing and reaction, meaning that all decisions had to be done by one and only one processing unit. And it worked. So why does distributed control appear?

Before we even start looking at the arguments it would be useful to define what author means with "distributed". Distributed in our case refers to physical structure of control system represented by separated processors which could work as a whole coordinating each other and cooperating. The most comprehensive system of such kind is human brain and the most known human constructed system of such kind is World Wide Web.

2.1 Team Cooperation and Education

The first and the most influential reason grows from competition characteristics and here consulting the competition site [2] for the rules and general idea is suggested to help understand the argumentation.

Being a part of a team revealed a lot of trouble young engineers have when for the first time they get in touch with a not so simple integrated system design. Less experienced students simply don't know how to start and thus need a strong supervisor to organize their work for them, more experienced students show tendency of doing everything alone. The general idea that is provided by Eurobot is for participants to experience team work and cooperation. None of the above natural tendencies, demonstrated by the students could possibly lead to fully comply with this brilliant idea in a half-year project. At least author's experience doesn't include positive results of such behavior.

The idea that could possibly save the situation lies, as usual, in between. For those who start over we have to provide a general plan, that will work but still give them enough space to make their own decisions. The plan should be supported with general equipment that could be expanded by the team members to current project needs. And decision making should allow mistakes to learn from. If we take a half-year period and the real amount of work to be done to successfully finish the project it becomes obvious that the whole scheme with freedom should be well thought of.

Suggested distributed approach steps towards the scheme which will help new comers understand the value of team work and to construct a working solution to the rules of the competitions on a team basis.

Speaking of teams from 4 to 7 people Eurobot provides at least 7 subsystems[1] to be realized. That means that each team member could possibly design at least one subsystem. It would be natural to make the process of design and construction concurrent where possible to benefit from team work though parts of final integrated design should be of course coordinated from the very start. High quality coordination requirements lead to general methodology of educational and commercial projects, which can be revised and improved each competition year.

Summarizing statements above, we come to a competition solution as defining separate work-parts of robot and giving those parts to team members for concurrent design. These parts include mechanics, electronics and programming. Design process of work-parts should be as independent from other parts as possible. By carrying out robotic project this way we benefit from team work and individual experience of team members. And in this case for a wide range of purposes (new comers, education, rapid design etc.) we need common hardware and software core which allows broad extension to current project needs on the educational level of the team member.

2.2   Effective System Resources Utilization

The question of natural resources preservation is becoming more and more actual as years pass and planet reserves are exhausting. Problems of human-designed system effectiveness are closely connected to this question.

Main resource that is consumed onboard of any mobile robot today is electric power. Generally all mobile robots use batteries of different kinds to work autonomously. As battery capacity is usually perceptibly limited question of power consumption has to be accurately considered when designing the system. As a consequence this leads to overall better effectiveness in terms of power consumption per action when compared to stationary power supplied systems.

To understand the order of consumption difference we'll compare an average EPIC standard embedded PC board with a Celeron M processor at 600MHz as the most powerful control solution, an average PCI104 standard embedded PC board with Geode LX 800 processor at 500MHz and PC104 standard embedded PC board with Vortex86DX processor at 800MHz as an intermediate control solution, ARM-core based microcontroller board at 36MHz as an advanced distributed control solution, and AVR-core based microcontroller board at 16MHz as a simple distributed control

---

[1] This article is not aimed at precise system design description and thus the number given is mainly for reference purposes. Still the number was taken from a real project of Eurobot 2010 carried out by beArobot team.

solution. The next table gathers all information needed for analysis at one place. Again this analysis is mostly approximation of the real system, showing only order of difference between radical solutions. Time lasting estimation is done using typical battery capacity of 2100mAh and with assumption that processing unit is working alone on same battery.

Table 1. Data processing unit's power consumption comparison in mobile robot application

Architecture	Manufacturer info	Power	Lasting time
Celeron M @ 600MHz	12V x 1.7A	20.4W	1h15m
Geode LX 800 @ 500MHz	5V x 1.3A	6.5W	3h50m
Vortex86DX @ 800MHz	5V x 0.74A	3.7W	6h50m
ARM-core proc @ 36MHz	11.1V x 0.16A	1.776W	14h10m
AVR-core proc @ 16MHz	11.1V x 0.09A	0.999W	25h15m

As you can see the better the calculation speed is the more power is consumed. It is to be noted that large power consumption for PC-compatible processing units would usually be spent less effectively because of different types of software latencies. Moreover if application is demanding real-time processing as most of robotic applications do software latencies are becoming a crucial point in system design. As microcontroller software is simpler and "closer" to hardware it is much easier to realize a real-time application with it.

If we do some search on distributed systems research by other people we can find useful confirmation [3] of above assumptions: "Except for simple microcontrollers with low performance and only basic features, modern processors exhibit serious drawbacks when employed in embedded real-time applications. ...No high performance processor enables the exact calculation of a program's execution time even at the machine code level, as is possible for the 8-bit microprocessors." That's why any PC-compatible board must be accompanied by at least one microcontroller in a robotics application.

Complicated tasks in a robotic application that will require high performance processor will be: vision – image processing; and tactics[2] – large amount of data processing. These tasks unlike others (like navigation or even more strict sensor calculation) do not demand strict real-time and thus can be realized on a general high performance PC. But again if we look at power consumption of such PCs like Celeron M we can see that equivalent network of microcontrollers could count from 20 pieces of solely AVR-core based (equivalent frequency 320MHz) to 11 pieces of solely ARM-core based (equivalent frequency 396MHz).

If we take the most consuming choice as a reference for the whole system top allowable power consumption for data processing purposes we could build two possible types of distributed systems instead:

---

[2] Strategy questions could be even more demanding, but we are staying in Eurobot competition limits, assuming strategy is formed by people and programmed statically. Tactics in this case can be thought of as a number of current preprogrammed actions put together dynamically by the robot.

- Main layer consisting of AVR- or ARM-core microcontrollers and one dedicated Geode LX 800 or Vortex86DX compliant processor for high demanding, complicated tasks;
- Network of devices based only on AVR- or ARM-core microcontrollers.

Both types are able to solve all tasks described but offer better scalability and power consumption quantization. Both leading to overall better system effectiveness and better correspondence to a definite real-time system task. Naturally compared to a centralized serial processing this approach will lead to equal or better performance and to longer battery life in general.

2.3 Cost Effectiveness

Questions of costs usually arise when project resources are limited which is true for all student Eurobot projects the author was involved in. First thing to mention here is that robot project no matter how simple the final solution will be demands a lot of resources which could be usually converted into money costs or time-to-produce equivalent for general model presentation.

Resources needed to build a robot include mechanical parts, electronics of different sort and software depending on chosen electronics. When designing a new robot each team has opportunities to do everything by itself or partially order parts from third parties. Question whether to use third party parts or not is driven by quality characteristics (whether team members can achieve needed by the project level of production quality) or by time characteristics (whether team members can do all work needed for the project in time). And of course project's budget should also be considered when deciding on such questions.

The aim of this article is not to propose a detailed model of costs in a robotic project, so we are going to analyze only the influence of different control system solutions based on the previous subsection. And again, this analysis is mostly approximation of the real system, showing only order of difference between radical solutions.

Table 2. Data processing unit's cost for mobile robot project[3]

Architecture	Equivalent relative unit cost
Celeron M @ 600MHz	7.7
Geode LX 800 @ 500MHz	5.7
Vortex86DX @ 800MHz	5.7
ARM-core proc @ 36MHz	1.3
AVR-core proc @ 16MHz	1

Above table speaks for itself – simple AVR- or ARM-core solutions proposed for distributed control system design are much cheaper than other options. Other than that they can be designed and produced by the team itself. This is not true for other options which are much more complicated in design and cannot be realized by the team in Eurobot project time period.

---

[3] For obvious reasons we cannot discuss current absolute costs. Author believes relative costs to be more accurate with time and containing more information. Base values for presented costs are actual for February 2010, Russia, Moscow.

Another point concerns reliability and parts replacement in case of malfunction. It is a usual practice to have spares for robot's important parts. In case of a single Celeron M solution we'll have to pay a lot more to have a part which we might not even use if all goes well. Almost the same situation is happening if we use Geode LX 800 or Vortex86DX in the design.

Positive outcome of a distributed system consisting of simple identical parts allows us not to have spares for all of them and thus reduce costs for control system. Quantization of total cost in this case allows us to assemble control system closely related to project budget. Possibility of self-development of distributed parts leads to engineer experience for team members and even more reduction in total costs.

2.4 General Software and Hardware Simplification

In suggested distributed system we can put less effort in complexity of separate system parts. Most picturesque analogue is human brain with single neuron as a basic part of control system. It is believed that we know the exact mechanism of neuron functioning [4]. Simplicity of each neuron is determined by its activation function, hiding system complexity in numbers of neurons and their interconnections.

Though full brain function yet cannot be reproduced in artificial design we can benefit from learning how Nature organizes complex systems. According to H. Simon [1] each complex system is hierarchical in structure. Moreover because of such hierarchical structures we are able to learn how world is organized as all of hierarchies possess similar properties independent of current realization. To reveal these similarities Simon suggests paying more attention to intensity of interaction between system parts rather than any other property (like spatial location of parts for example). Hierarchy levels are actually intermediate stable forms of systems on their evolution way. And speaking of interaction it can be observed inside a level between its parts and also between levels which is the most interesting part for a researcher.

Such argumentation could strengthen and help explain the idea of simplifying system parts and arranging them in interconnected nets as it reflects Natural tendencies. Simplicity will influence individual developers on the team giving them more freedom in work. To achieve that, special layer of software should be added to the system to glue all parts of the final project together. This layer has to organize communication inside the whole configuration and is the most influential part of it. That is why it has to be developed with much care making it the most difficult part of system software on the robot. But once realized it can be reused in different projects. Moreover it can become a part of educational methodology for young developers restricting their style of programming.

If there is a question of simplification, education and general usage involved we should pay more attention to how the system is realized. In such cases we should think even more than usually of such a design that will help its user in his work. Distributed system we are discussing here contains software and hardware elements. The user being interested in solving a number of tasks put for him by the competition rules should have easy and clear access to system functionality. Speaking the words of D. Norman [5] a good design should comply with two fundamental principles:

- Clear conceptual model;
- Visualization.

Using these principles and other design techniques is important on all stages of development. Design defines how easy it is to work with the system and thus success of the principle, current project and future projects which might be based on current one.

Summarizing main ideas of subsection it is important to highlight possibility of simplifying a system design by introducing separation to control parts as it is done in advanced Natural control formations like human brain. By making each part of the system simple we develop a hierarchical system structure with two main layers. One consisting of parts themselves and the other consisting of interconnections and communication means. And if the first one is represented mostly by hardware parts, then the second demands accurate universal software core. To be successful both levels must be designed easy to use.

2.5 Robust Design

When we speak about mobile objects like cars or robots we think of highly reliable devices able to operate even in harsh environments. Usually to achieve such reliability parameters of systems are highly reserved. In case of electronic control system it is a good practice to introduce duplication into design configuration. Robust in this case would mean able to work when some random parts of the system are not working right.

First question of robustness reflects its essence. There should be a mechanism in a system to continue operation while some parts show malfunction. Solution to this question lies on hardware and software levels. On hardware level there should be a physical way to reallocate control lines from a broken part to a working part. On software level there should be a mechanism to determine which part is broken if broken at all and reallocate its control functions to other working parts.

It is known that human brain is highly robust and mechanisms used for communication between neurons allow quick reorganization for alternate routes. When operating as usual these mechanisms allow load balancing and in case of emergency they ensure save of functions. Artificial robust control system should possess similar properties, the most important question being physical alternate interconnection route organization techniques.

Second question of robustness is how quickly system can be repaired in case of malfunction of some or all parts. This question is highly connected to the one discussed earlier about spare parts.

Distributed system naturally will have more potential to develop robust techniques rather than highly concentrated system. Mainly because questions of robustness are similar to those questions vital for distributed system's existence.

2.6 Research Platform

Inventions of the past allow us to progress to better solutions of the future. Ability to discover does not belong exclusively to fundamental sciences. If we refer to H. Simon we'll see that he considered design engineering a special branch of activity different from natural or human sciences and different first of all in methods of progressing. In his book [1] Simon takes a step to formulate what should sciences of artificial teach and what to include. And once over with his book it becomes obvious that sciences of artificial or design engineering mainly use synthesis rather than analysis. All tools and methods supplied to young engineers in universities should mainly concentrate on design techniques and they require cardinally different, special logic. Before Simon science was seen from such a perspective that artificial was a product of science and there is nothing to investigate, to research in it. On his behalf Simon highlighted the meaning and presence of artificial in Nature.

Research in engineering means a special strategy of search by synthesis. It's much of a trial and error strategy. And it can be explained again with brain analogy. When we cannot analytically come to understanding how Natural system works we have nothing to do but to try reproducing it by synthesis of known elements. We know how neuron works but we don't know how a net of neurons gains ability to intelligent behavior. We could try to engineer a system with similar behavior and discover yet unknown correlations which could bring us to a new level of understanding of how Natural process is organized.

Distributed system being described could become such a research platform. Distributed parts staying physically unchanged could serve as an inexhaustible source for software development and interconnection organization.

In spite of the wish to simplify system parts it would be right to address another scientist, John von Neumann. In his unfinished work [6] von Neumann searched answers for two main questions which could possibly be asked later in our research work:

- How to construct a reliable system out of unreliable parts?
- What logic organization should be sufficient for a system to reproduce itself?

John von Neumann believed that when you start to work in a new scientific field it is right to start with problems that could be clearly formulated, even if they refer to ordinary effects and lead to good known results. Later a strict theory explaining such results could become the foundation of progress.

Proposed distributed system could possibly carry no unknown elements in terms of nowadays' electronics or software engineering but it carries the potential to discover.

3   System Structure, Principles and Realization

To start describing how distributed system is meant to be organized let's briefly summarize background and ideas that actually influence the future configuration (you can find more accurate argumentation in the previous section). These points describe goals to realize in proposed system and consequences of such realizations:

- Simplification of team work – method to organize cooperation of developers;
- Simplification of system hardware – capabilities to self-development of system hardware and education capabilities;
- User friendly design – more system users, general design success;
- Introduction of basic control modules – modular design method, capabilities of rapid development;
- Cheapening system elements – more accurate quantization of system costs;

- Effective resource utilization – more accurate quantization of system power and performance;
- Foundation of future research platform – base unchanged hardware allowing easier and quicker research in interconnection and communication techniques, robustness and software control.

Now we know exactly what we want to achieve with the new system and it's time to discuss the means and methods. In other words we must understand how those goals can be accomplished. For now it should be clear that we are discussing a control system mostly to be realized in hardware and this system has to be distributed in its core organization.

Main system structure and at the same time usage diagram of distributed architecture is presented in figure 1. Schematics is meant to be rather straight forward to understand most of ideas from the figure itself but there are several important remarks to make.

It would be good to have a universal control system at the end of development as success of any design is determined by the number of uses and users. But it should be noted that current version is developed in spite of Eurobot competitions and thus reflect the structure of challenges typical for competition rules. This, of course, doesn't necessarily mean that the system cannot be used in other applications; it means that some parts of the system should be considered as universal and others as application specific.

If we take a look at figure 1 we'll see that there are 7 embedded control boards each on a specific system level. Levels should be considered universal and thus present current distributed control system architecture understanding. Any mobile robot control application is meant to be same in level structure.

On the other hand total number of embedded boards and their number on each level are considered to be application specific. Generally the number of boards is equal to the number of system tasks.

If we take Eurobot match as an example we'll see that first of all robot must be able to move ("Navigation acting level") – this time we use a simple differential drive for that purpose requiring only one control unit as there is not much data from sensors to be processed.

Secondly there are always a number of mechanics oriented tasks to realize. For example: robot should gather balls to some storage inside its body and unload them when needed. Embedded boards to realize control for such tasks are located on "General task level". And this time there are 3 tasks, each with unique mechanical background, sensors and processing, and 3 control boards for them.

Thirdly there are questions of navigation which refer to localization of robot itself and objects in its environment. Questions from one side include relative positioning techniques which allow gathering information relative to current robot position, tracking robot movement and presence of stationary obstacles on the way. On the other side, accurate localization of robot and its opponent in the environment. These tasks are represented with 2 control boards in the model structure.

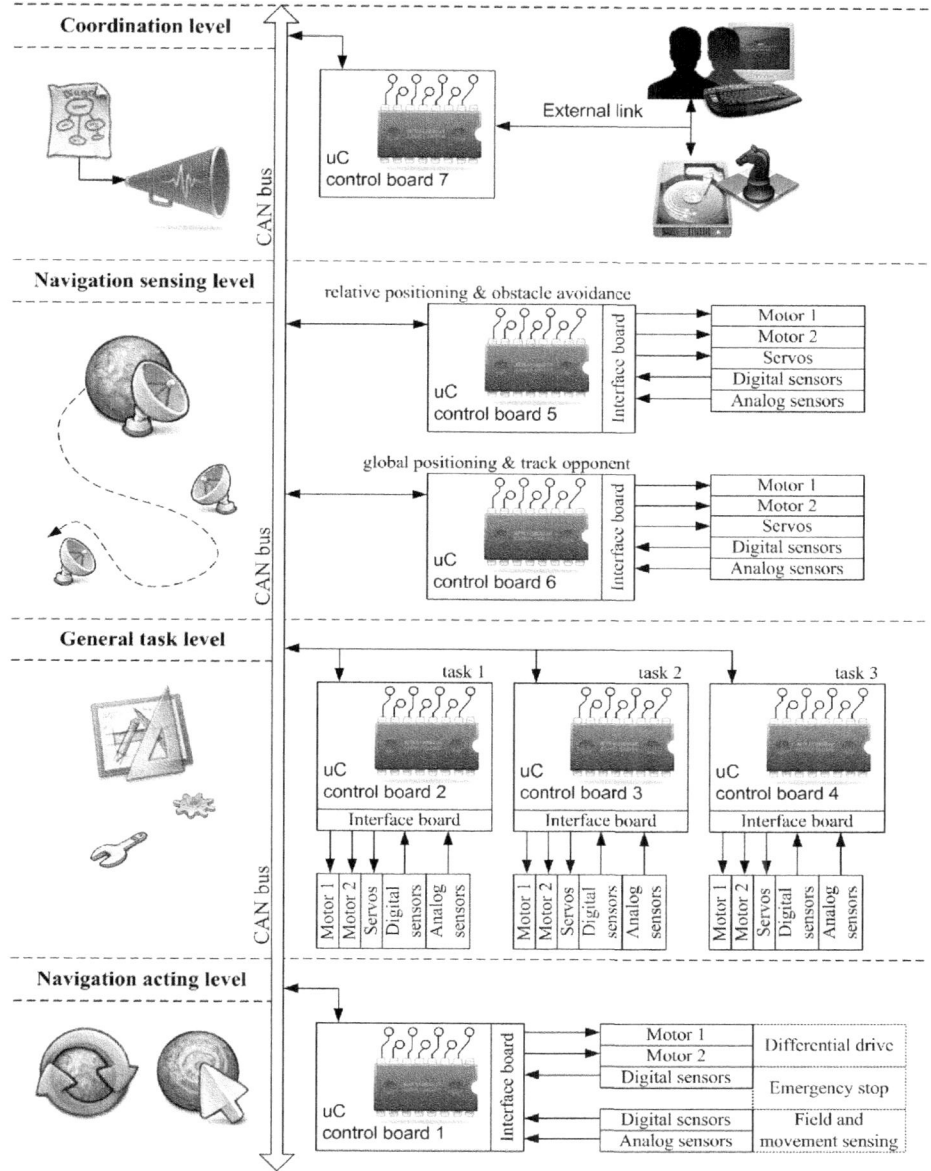

Fig. 1. Structure of mobile robot distributed control system

At last there is a level that represents the whole distributed system – "Coordination level". Control boards belonging to it contain software core needed to start the system, possibly support its functioning while system is active, they provide system with additional calculation power and memory to share and finally they form a gateway to out-of-system control and communication. If you lack resources for some task this is the first place to consider additional board modules. Current system architecture is similar to asymmetrical multiprocessor architecture reviewed by M. Colnarič and others [3].

When designing a control system of mobile robot there is a lot of debug work for developers especially in programming field. Taking into account possible amount of embedded devices normally we should carefully choose their placement onboard the robot. Decision depends on easy access for debug and embedded program uploading on one hand and position close to controlled hardware for less noise emission and reception on the other.

Figure 2 helps to understand how hardware installation and reprogramming is simplified by putting all hardware descriptions in a single configuration profile of the system. We are not considering software core organization this time, but for configuration profile idea to work we should program devices on all architecture levels except coordination a special but identical software code. This startup code should be capable of receiving setup commands from coordinator and reprogram device functions if needed. Once online all devices can be configured to current system application profile by the leader on coordination level.

First step to realization of introduced system architecture is presented on figure 3. It's an example of control board developed by beArobot team for Eurobot 2010 competitions. We won't spend much time on describing all properties of the board as it would be enough information for a separate article. But what is to be highlighted is the need for a special interface board that is one of the key points of introduced architecture. It allows changing to be done only in the interface, using same control board, for all structural levels of the system and it gives capabilities to teach printed board design and circuit engineering without significant revision of the whole system mainly on the basis of mechanics tasks as they change each year of competitions.

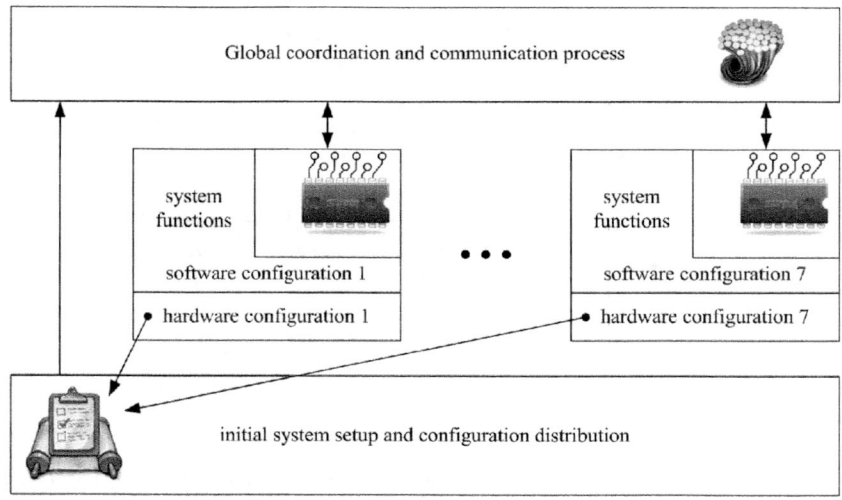

Fig. 2. General idea of configuring distributed control system

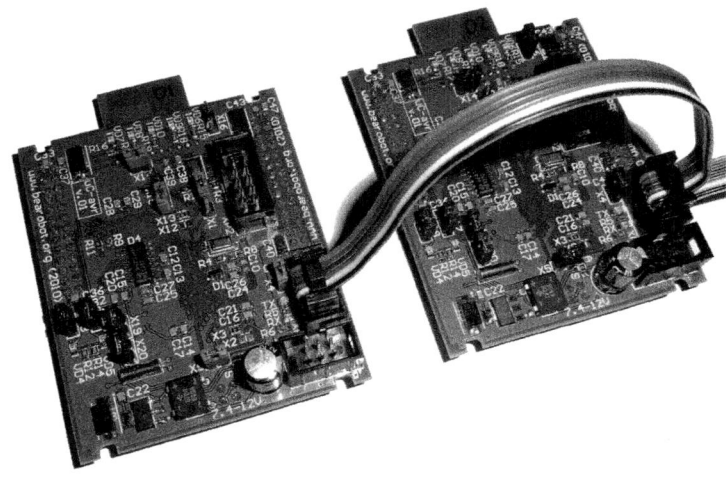

Fig. 3. Hardware control units – interconnected embedded boards example

Boards are based on AVR-core microcontroller and offer a wide range of communication capabilities which can be easily accessed with a number of connectors. No task specific parts are included onboard. Most of microcontroller ports' pins are traced to two special connectors meant to be standard for all other control boards with different core microcontroller. This standard allows a number of functional boards to be developed by team members while working on another project or use functional boards developed for other projects as well. Special effort was put in hardware configuration abilities of the boards with jumpers and soldered components, so that most of pins naturally used for onboard needs can be also accessed within standard connectors.

Optionally and mainly for debug purposes each board has a wireless connection capability. This is done to help developers have easy software access to each board without plugging in any cables. That could be useful whenever the whole system is installed on the robot but some debug processes are still carried out on individual devices. In normal operation wireless connectivity can be switched off by hardware or software means.

4   Conclusions

This article presents the first step towards universal distributed control system for mobile robot applications. In the first place author expects this solution to be useful for student teams studying through Eurobot competitions. One of the far going goals and usages for the architecture scarcely described is forming of practical training methodology for future engineers.

Work on the subject is far from completion. Still, provided argumentation allows further development and ideas posited propose future research as a natural application of the system. Moreover if you look back at the beginning you'll see that research was the cause of even starting this work. But unfortunately we are still several steps before that main research itself. For now some auxiliary work is necessary to advance to outskirts of control science.

As the article is written while preparing for another Eurobot competition author expects to use results of this work in a real application to test ideas, to get solutions and to find improvements. Natural result of such work would be software and hardware realization which could become a subject of another article.

References

1. Simon, H.: The sciences of the artificial, 3rd edn. Massachusetts Institute of Technology, Cambridge (1996)
2. Eurobot, international robotics contest, http://www.eurobot.org
3. Colnarič, M., Verber, D., Halang, W.A.: Distributed Embedded Control Systems. In: Advances in Industrial Control Series, pp. 62–63. Springer-Verlag London Limited, Heidelberg (2008)
4. McCulloch, W.S., Pitts, W.: A Logical Calculus of the Ideas Immanent in Nervous Activity. Bulletin of Mathematical Biophysics 5, 115–133 (1943)
5. Norman, D.A.: The Design of Everyday Things. Basic Books, New York (2002)
6. Neumann, J.: Theory of Self-Reproducing Automata, 2nd edn. Edited and completed by Burks. A.W. University of Illinois Press, Champaign (1966)
7. Penrose, R.: The Emperor's New Mind: Concerning Computers, Minds, and The Laws of Physics. Oxford University Press, New York (1999)

A Modular and Scalable Electronic System for Mobile and Autonomous Robots

Raimund Edlinger and Michael Zauner

Upper Austria University of Applied Sciences,
Research & Development,
Stelzhamerstrasse 23, 4600 Wels, Austria
{raimund.edlinger,michael.zauner}@fh-wels.at
http://www.fh-ooe.at/en/rd/forschung/

Abstract. Robotics is a scientific discipline which includes the operation and development of robots and other fields of computer science, as well as electrical and mechanical engineering and high-performance modular control systems. This paper reports the first results of a research project that uses a previously developed *MES - Modular Electronic System* for autonomous mobile robots and for developing robots to use the MES in other scientific applications. This approach offers the ability to build a scalable and flexible control system with only a few modules. A novel feature of our approach is its modular integration of electronics with the ability to multitask, which can be adapted for particular applications. We demonstrate the robustness of this approach in controlling an indoor mobile robot for the *EUROBOT 2009 and 2010*.

Keywords: robotic hardware and software architectures, autonomous robots, controlling system, multitasking architecture.

1   Introduction

Robotics is a scientific discipline which includes the operation and development of robots and other fields of computer science, as well as electrical and mechanical engineering. This requires a high-performance control system that is divided into a *Modular Electronic System (MES)*. In this paper we present an approach for robot development using the MES in various applications. Put simply, the MES consists of a multi-processor system which allocates tasks to different modules. In rough terrain, for example that in which rescue robots are commonly employed, the MES can be implemented, as well as in competitions where the playing field has been predefined, such as the RoboCup Soccer-Leagues for smaller robots or the EUROBOT. This system deals with the challenge of combining an efficient, flexible and scalable control system with small dimensions.

2   Related Work

Much of the recent work in robotics has used embedded systems such as PC 104 [10], Mini-ITX [9] or the RNBFRA-Board [8]. In contrast to the proposed

D. Obdržálek and A. Gottscheber (Eds.): EUROBOT 2010, CCIS 156, pp. 193–200, 2011.

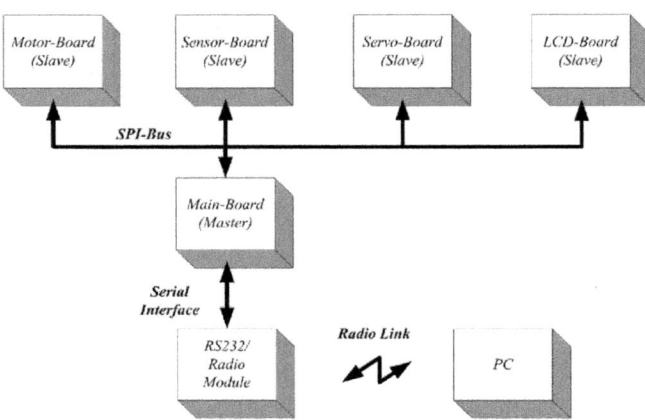

Fig. 1. Modular electronic system

concept, these systems are rather inflexible and cannot be easily adapted to meet specific needs. The approach presented herein is advantageous, because the modular control system can be placed nearer to the sensor or actuator. This makes it well suited for driving or grasping tasks, and moreover reduces the length of the cables. In particular, due to sensor disturbance, it is often necessary to separate the power devices spatially from sensors.

An important consideration in the design of such robots is the cable connectors. The main board forms the central unit of the modular system. In the first generation it was possible to connect only 8 slaves. However, in high level applications this is often not enough. Hence, in the new version it is possible to connect 14 interfaces for different slaves, see figure 1. Due to space considerations, in the current design the plugs were changed to ensure a better hold and because of their smaller dimensions. For the new system plugs from Micro-MaTch [11] are used.

3   Hardware Design

On each of these boards the micro controller ATMega644 [7] is used. The main board (master) distributes tasks to the slaves (sensor, servo and motor board) and the system is capable of processing them simultaneously. The slave boards are connected via the SPI[1] bus to the master board. The SPI allows for a high speed connection between master and the slaves boards. The data is transfered at a rate of 625 kBaud. The master transfers the data via the MOSI pin to the slaves and via the MISO pin from slave to master, see figure 2.

Figure 3 shows that the master is connected with all slaves via the MISO, MOSI and SCK piplines. Each of the 14 SPI plugs is given a unique identifier. By means of this connection the single modules are activated, enabling data transfer.

---

[1] Serial Peripheral Interface.

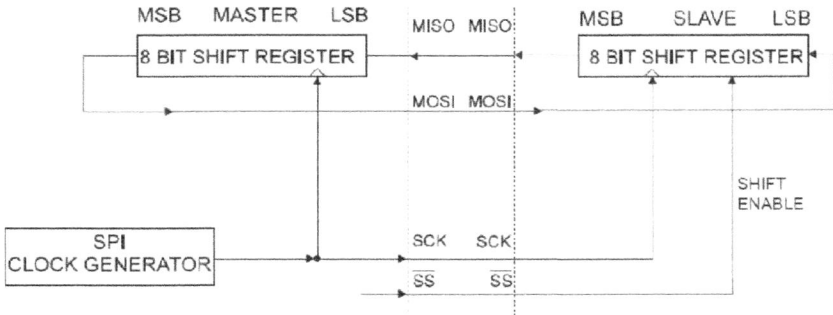

Fig. 2. Function of the SPI bus

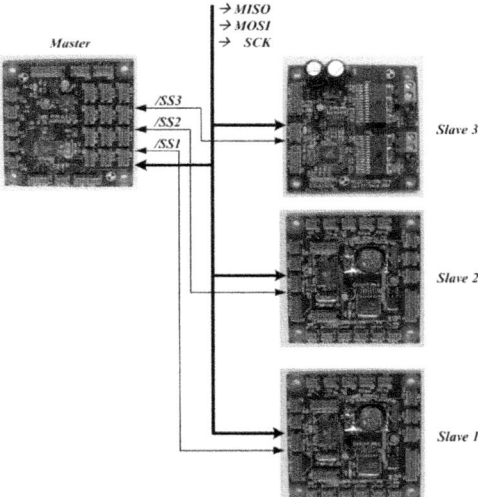

Fig. 3. Block diagram of SPI

3.1 Master Board

Main Board. The main board distributes the tasks to the slave boards. It is the central unit which communicates with the slaves via the SPI. It is possible to connect up to 14 slaves to the main board, up to five digital or analogue I/Os and the communication interface, see figure 4. The main board executes the main program which includes all the strategies, the timing, the route planning, etc.

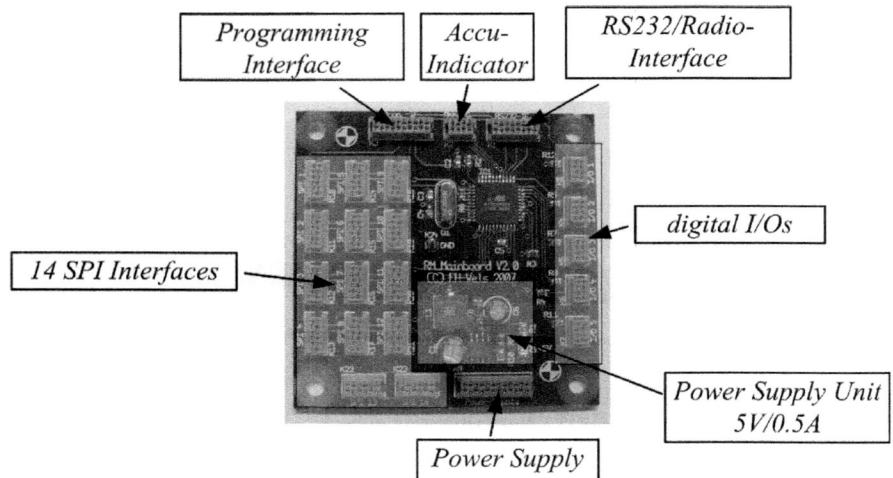

Fig. 4. Mainboard

The board includes the following interfaces:

- 14 SPI interfaces and five analog or digital I/Os
- a control connector for accumulator tension
- a connection to the universal communication board and it is able to choose between the RS232 interfaces, radio interface and serial interface with TTL[2]
- supply interface
- programming interface

The voltage regulator *LM2594* [5] from *National Semiconductor* [4] supplies the electronic and the periphery with 5 V. With this switch regulator it is possible to provide a maximum current of 500 mA.

3.2 Slave Boards

Sensor and Servo Board. Figure 5 represents a PCB[3] to use it for more applications. On the one hand it is possible to connect up to 16 sensors (sonar, infrared, etc.). On the other hand this board can control up to 16 servos, e.g. Hitec [12], Futaba [13], etc. The sensor and servo board in figure 5 includes the following interfaces:

- communication connector with the master via SPI
- supply interface
- programming interface
- and 16 sensor or servo interfaces

The switch regulator LM2677 [6] provides the electronic and the periphery with 5 V. With this switch regulator a maximum current of 5 A is possible.

---

[2] Transistor Transistor Logic.
[3] Printed Circuit Board.

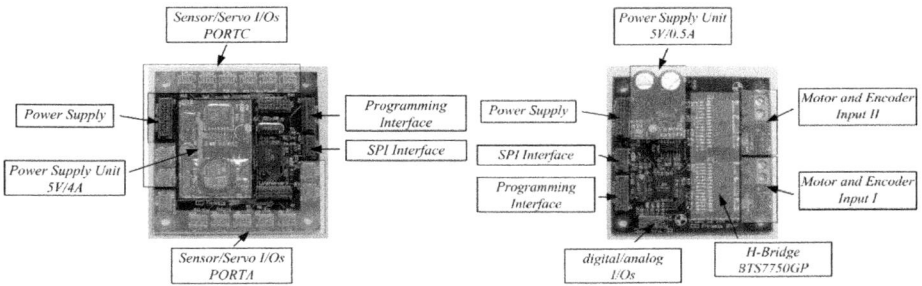

Fig. 5. Servo/Sensor board and Motorboard

Motor Board. The motor board has two full H-Bridges BTS7750 [14] from Infineon to control two DC motors or one stepper motor, see figure 5. The BTS 7750 GP is part of the TrilithIC family and has the following features:

– low R_{DSon}: 70 mΩ high-side switch, 45 mΩ low-side switch
– maximum peak current of 12 A
– full short-circuit protection
– operates up to 40 V
– PWM frequencies up to 1 kHz

It controls the speed, the position and the odometric navigation system.

Odometric Navigation. The basic navigation system is based on odometric navigation. As shown in figure 6 the robot will measure the increments of the left and the right wheel cyclically and calculate the differential path ds and the angle φ_R, see equations 1 to 8. With these variables the exact x and y-position can be calculated. At the beginning the x and y coordinates and the starting angle must be known and these values are stored in the micro controller.

$$s_n = \frac{1}{2}(Increments_{right} + Increments_{left})k_S \qquad (1)$$

$$\varphi_R = \varphi_0 + (Increments_{right} - Increments_{left})k_D \qquad (2)$$

$$ds = s_n - s_{n-1} \qquad (3)$$

$$x_R = x_{R-1} + ds\cos\varphi_R \qquad (4)$$

$$y_R = y_{R-1} + ds\sin\varphi_R \qquad (5)$$

It is possible to use general method, see figure 7, though kinematic equations, for evaluate the correct robots position and orientation:

$$l = \sqrt{\Delta x^2 + \Delta y^2} \qquad (6)$$

$$l = \sqrt{(x_A - x_B)^2 + (y_A - y_B)^2} \qquad (7)$$

$$\varphi = \varphi' - \arcsin(\frac{\Delta y}{l}) \qquad (8)$$

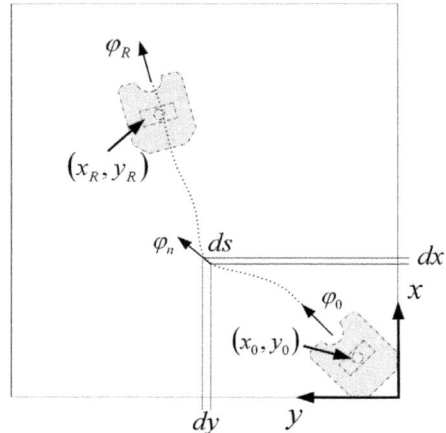

Fig. 6. Odometric Navigation

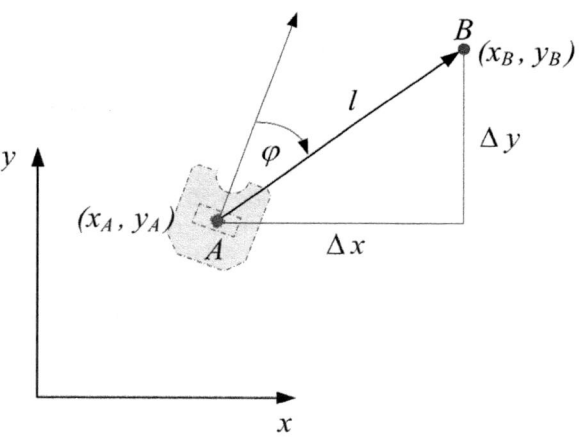

Fig. 7. Path Planning

Communication module. The communication board has three different ports. Firstly it can communicate via a RS232 interface. Secondly it can correspond via a radio module and thirdly it can talk over a TTL interface.

USB to Radio Module. This module is the connector between the robot and the debugging computer. It enables the data to be easily read out of the robot (without a cable).

4   Multitasking System

In the modular system of the first generation the duties of the man board became in one State-Machine exhausted. Only one job per time thereby was able to finish (driving mode or another task). This fact has some disadvantages for the performance of the robot. Hence, it was a big aim for the modular system of the second generation to implement a high performance and self developed cooperative multitasking system [2].

5   Experimental Results

For the development of the Eurobot robot 2009 makes use of one main board, one motor board, one communication module, four sensor/servo boards, one LCD module, one programming interface and one power supply PCB.

The Eurobot robot in figure 8 is divided into two zones. At the front are two grasping systems to handle the column elements. On the top of the robot there are storage units, so two columns can be placed there. That way the robot can transport four columns, two in the storage units and two using the grasping system. On the back is one grasping system to handle the lintels. The robot can transport two lintels, one in an additional storage unit and one with the grasping system. The robot has 17 ultrasonic sensors to detect obstacles and temples. The control unit is a modular electronic system, that is supplied with 14.8V. The opponent robot detection system, the programming interface and the emergency stop switch are located on top of the robot.

Fig. 8. Robot for Temples of Atlantis

6   Conclusion

The modular electronic system for the construction of autonomous robots exists of three different basic modules (main, servo and sensor modules) and two additional modules (communication and LCD module, which is not introduced). The additional modules allow to realize an easy human machine interface. With this system it is possible to do processes however and it becomes a better management and high performance to process tasks at the same time. Easy status announcements are able to send from the radio module to an operator or the robot can become remotely controlled with easy orders. The LCD module enables the user to carry out changes in the robot, without changing the program and so a connection to an operator station is not necessary. Moreover, important information about the state of the robot is displayed on this module. The introduced system offers an amount in possibilities to build robots in short time with wide activity field offers for adaptations.

Acknowledgements. The authors grateful acknowledge the support of the Upper Austria University of Applied Sciences , National Instruments, Elra Motors, Beta Layout and SMC for setting up the Robo Racing Team.

References

1. Brandsteidl, B., Edlinger, R.: Modulares elektronisches System. Technical paper (2007)
2. Bräunl, Thomas, Embedded Robotics. Springer, Heidelberg (2003)
3. Haiduc, Pavel, Hilfe für CodeVisionAVR C-Compiler, HP InfoTech s (2005)
4. National Semiconductor (2008), http://www.national.com/analog/power
5. National Semiconductor LM2594 (2002), http://www.national.com/analog/power
6. National Semiconductor LM2677 (2008), http://www.national.com/analog/power
7. Atmel (2006), http://www.atmel.com/products/avr/default.asp
8. RoboterNETZ (2010), http://rn-wissen.de/index.php/RNBFRA-Board
9. Mini-ITX (2010), http://de.wikipedia.org/wiki/Mini-ITX
10. PC104 (2010), http://www.pc104.org/
11. Micro-MaTch (2010), http://www.tycoelectronics.com/catalog/minf/en/439
12. Hitec (2010), http://www.hitecrc.de/store/home.php?cat=308
13. Futaba (2010), http://www.futaba-rc.com/servos/index.html
14. Infineon (2010), http://www.infineon.com/cms/en/product/findProductTypeByName.html?q=BTS7750

Author Index

Barilla-Pérez, María Enriqueta   1
Bittel, Oliver   72
Blaich, Michael   72
Borovik, Rustam   168

Chistyakov, Mikhail   168
Colín-Espinoza, Andrés   1
Cubek, Richard   118

Demidov, Andrey   168
Dessimoz, Jean-Daniel   14, 30

Edlinger, Raimund   46, 193
Enescu, Catalina   79
Ertel, Wolfgang   118

Faigl, Jan   93
Fichera, Loris   57
Fromm, Tobias   118

Gauggel, Dominik   72

Halgašík, Jaroslav   93
Hensler, Jens   72
Hirt, Julian   72

Ilas, Constantin   79, 108

Jamal, Rahman   87

Krajník, Tomáš   93
Krasnobryzhiy, Boris   168
Kuturov, Andrey   168

Marletta, Daniele   57
Montes-Venegas, Héctor Alejandro   1
Morega, Alexandru   79
Mudrová, Lenka   93

Nicosia, Vincenzo   57
Novischi, Dan   108

Paczynski, Andreas   130
Paturca, Sanda   79, 108
Pölzleithner, Andreas   46

Santoro, Corrado   57
Schneider, Markus   118
Stetter, Ralf   130
Sukhotskiy, Dmitry   141

Vlasov, Andrey   180

Wilcox, Tony   149

Yudin, Anton   141, 168, 180

Zauner, Michael   46, 193
Ziemniak, Paweł   130

GPSR Compliance

The European Union's (EU) General Product Safety Regulation (GPSR)
is a set of rules that requires consumer products to be safe and our
obligations to ensure this.

If you have any concerns about our products, you can contact us on
ProductSafety@springernature.com

In case Publisher is established outside the EU, the EU authorized
representative is:

Springer Nature Customer Service Center GmbH
Europaplatz 3
69115 Heidelberg, Germany

Batch number: 09490872

Printed by Printforce, the Netherlands